JOINT NATURE CONSERVATION COMMITTEE

Vegetation communities of British lakes:
a revised classification

Catherine Duigan,
Countryside Council for Wales,
Penrhos Road, Bangor,
Gwynedd, LL57 2DW

Warren Kovach,
Kovach Computing Services,
85 Nant Y Felin, Pentraeth,
Anglesey, LL75 8UY

Margaret Palmer,
Nethercott, Stamford Road,
Barnack, Stamford,
PE9 3EZ

*Joint Nature Conservation Committee,
Monkstone House, City Road,
Peterborough, PE1 1JY*

Cyngor Cefn Gwlad Cymru
Countryside Council for Wales

SCOTTISH NATURAL HERITAGE

ENGLISH NATURE

defra
Department for Environment
Food and Rural Affairs

Llywodraeth Cynulliad Cymru
Welsh Assembly Government
CORFF NODDEDIG | SPONSORED BODY

SCOTTISH EXECUTIVE

1

Contents

List of tables and figures 5

Summary 6

Chapter 1   **Introduction** 7

Chapter 2   **Survey method** 9

Chapter 3   **The dataset, data analysis and the Plant Lake Ecotype Index (PLEX)** 10
             The dataset 10
             Data analysis 10
             Plant Lake Ecotype Index 12

Chapter 4   **Revised classification system and lake key** 14

Chapter 5   **Individual descriptions of the lake groups** 21
Group A   Small, predominantly northern dystrophic peat or heathland pools,
             dominated by *Sphagnum* spp. 23
Group B   Widespread, usually low-lying acid moorland or heathland pools and small lakes,
             with a limited range of plants, especially *Juncus bulbosus*, *Potamogeton
             polygonifolius* and *Sphagnum* spp. 24
Group C1   Northern, usually small to medium-sized, acid, largely mountain lakes, with a
             limited range of plants, but *Juncus bulbosus* and *Sparganium angustifolium* constant 25
Group C2   North western, predominantly large, slightly acid, upland lakes, supporting a
             diversity of plant species, *Juncus bulbosus* constant, often with *Littorella uniflora*
             and *Lobelia dortmanna*, in association with *Myriophyllum alterniflorum* 26
Group D   Widespread, often large, mid-altitude circumneutral lakes, with a high diversity
             of plants, including *Littorella uniflora*, *Myriophyllum alterniflorum*, *Callitriche hamulata*,
             Fontinalis antipyretica and *Glyceria fluitans* 28
Group E   Northern, often large, low altitude and coastal, above-neutral lakes with high diversity
             of plant species, including *Littorella uniflora*, *Myriophyllum alterniflorum*, *Potamogeton
             perfoliatus* and *Chara* spp. 30
Group F   Widespread, usually medium-sized, lowland, above neutral lakes, with a limited
             range of species, but typified by water-lilies and other floating-leaved vegetation 32
Group G   Central and eastern, above neutral, lowland lakes, with *Lemna minor*, *Elodea
             canadensis*, *Potamogeton natans* and *Persicaria amphibia* 33
Group H   Northern, small, circumneutral, lowland lakes, with low species diversity
             characterised by the presence of *Glyceria fluitans* and *Callitriche stagnalis* 35
Group I   Widespread, mostly moderately large, base-rich lowland lakes, with *Chara* spp.,
             Myriophyllum spicatum and a diversity of *Potamogeton* species 36
Group J   Northern coastal, brackish lakes, with *Potamogeton pectinatus*, *Enteromorpha* spp.,
             Ruppia maritima and fucoid algae 38

Contents

Chapter 6   **Using the Plant Lake Ecotype Index (PLEX)** **45**
             Case Study One: Llangorse Lake, Wales (Lake Group I) 45
             Case Study Two: Loch Chon, Scotland (Lake Group C2) 47
             Conclusions 48

Chapter 7   **Discussion** **49**
             The lake resource 49
             Analysis and classification schemes 50
             Regional studies 52
             Future developments 53

Chapter 8   **Acknowledgements** **55**

Chapter 9   **References** **56**

Chapter 10 **Annexes** **58**
Annex A   A listing of submerged and floating macrophyte taxa included in the 1989 TWINSPAN
             analysis (Palmer *et al.* 1992) and the 2004 re-analysis of the JNCC lake dataset
             described in this report, with the number of records used 58
Annex B   Constancy table for standing water site types: submerged and floating plants
             with number of occurrences and PLEX values 61
Annex C   Relationships between the environmental variables, taxon richness and PLEX 64
Annex D   Sites included in this study 70

List of figures and tables

Page

Figure 3.1.   Distribution of lakes included in the analysis 10
Figure 3.2.   Dendrogram of TWINSPAN analysis showing major groups A-J 11
Figure 3.3.   Ranges of PLEX scores for lake groups 13
Figure 3.4.   PLEX scores plotted against pH 13
Figure 3.5.   PLEX scores plotted against alkalinity 13
Figure 4.1.   Ranges of altitude for lake groups 17
Figure 4.2.   Ranges of surface area for lake groups 17
Figure 4.3.   Ranges of pH for lake groups 17
Figure 4.4.   Ranges of conductivity for lake groups 17
Figure 4.5.   Ranges of alkalinity for lake groups 18
Figure 4.6.   Ranges of taxon richness for lake groups 18
Figure 4.7.   Number of sites per lake group for Scotland, England and Wales 18
Figure 4.8.   Ordination diagram of a CCA analysis, including the major environmental variables 19
Figure 6.1.   PLEX scores through time for Llangorse Lake 46
Figure 6.2.   PLEX scores through time for Loch Chon 47
Figure 7.1.   Distribution of standing waters in Britain 49
Figure 7.2.   Pie chart showing percentage distribution of lakes surveyed within groups 50
Figure 7.3.   An illustration of the relationship between the lake groups described in this report
              and those protected under the Habitats Directive 51

Table 4.1.   Constancy table for site groups 15
Table 4.2.   Key to standing water site types using submerged and floating taxa 16
Table 4.3.   Statistics for the CCA analysis in Figure 4.8 19
Table 4.4.   Statistics for the CCA analysis incorporating 100km grid squares 20
Table 5.1.   Rarity status of submerged and floating standing water plants 21
Table 6.1.   Calculating a time series of mean site PLEX scores for Llangorse Lake. 46
Table 6.2.   Calculating a time series of mean site PLEX scores for Loch Chon. 47
Table 7.1.   The relationship between Lake Groups in the new classification and Site Types in the previous
              classification: numbers of sites within each class. 51

Summary

The first comprehensive classification scheme for standing waters in Britain (Palmer 1992; Palmer *et al.* 1992) was based on macrophyte surveys carried out by the Nature Conservancy Council (NCC) from 1124 standing waters throughout England, Scotland and Wales between 1975 and 1988. This dataset became an important source of information used for a variety of purposes including conservation site selection and the identification of aquatic communities in the National Vegetation Classification (NVC) (Rodwell *et al.* 1995). The macrophyte records were also used in a number of atlases (Preston & Croft 1997; Preston *et al.* 2002).

Since then, the NCC and its successor organisations (Countryside Council for Wales (CCW), English Nature (EN) and Scottish Natural Heritage (SNH)) have commissioned a substantial number of additional lake surveys, leading to the establishment of a much larger dataset with records from 3447 sites (310 England, 38 Wales and 3099 Scotland). The advent of the Habitats Directive (*Council Directive 92/43/EEC on the conservation of natural habitats and of wild fauna and flora*) and more recently the Water Framework Directive (*Directive of the European Parliament and of the Council establishing a framework for Community action in the field of water policy*) provided the incentive for the production of a revised classification using this larger dataset, supplemented by environmental data.

This report presents the results of the statistical analysis of the enlarged dataset. Differences from the previous classification are highlighted. The important components of this report include:

◆ separate ecological descriptions of 11 distinct lake groups (A-J);

◆ a revised key for the classification of newly surveyed sites;

◆ updated distribution maps showing the distribution of sites within each lake group, supported by summary environmental data and macrophyte constancy tables;

◆ the earlier Trophic Ranking Scheme (TRS) is replaced by a Plant Lake Ecotype Index (PLEX), which reflects the fit with the new classification scheme rather than a perceived simple trophic relationship.

Chapter 1: Introduction

Aquatic plants are important ecological components of lakes, where they have a complex role in the structure and functioning of the ecosystem (Moss 1998; Scheffer 1998; Wetzel 2001; Pokorný & Květ 2004). The statutory conservation agencies in England, Wales and Scotland have a long history of carrying out routine macrophyte surveys of standing waters, which has led to the accumulation of a large dataset held by the Joint Nature Conservation Committee (JNCC). The primary aim of this survey effort was to describe the botanical resource of standing waters in Britain. Rare species and plant assemblages representative of lake types have been selected as features for protection on Sites of Special Scientific Interest (SSSIs) throughout this region. The process of conservation site selection requires the ability to compare sites and set them in a local and national context (Nature Conservancy Council 1989). For this reason, a national classification scheme for lakes was required.

Between 1975 and 1988, the Nature Conservancy Council (NCC) carried out surveys of macrophytes from 1124 sites throughout England, Scotland and Wales. The information gained from these surveys was used to develop the first comprehensive classification scheme for standing waters in Britain (Palmer 1992; Palmer *et al.* 1992). At that stage a possible bias was recognised towards sites that were likely to be botanically rich, or to have some other value for conservation (Palmer *et al.* 1992) and the majority of sites were in Scotland and northern England but natural lakes are found at relatively high densities in these areas. This dataset was an important source of information used for the identification of aquatic communities in the National Vegetation Classification (NVC) (Rodwell *et al.* 1995). The plant records were also incorporated in distribution maps in the account of the aquatic plants of Britain and Ireland (Preston & Croft 1997) and in the new atlas of vascular plants of Britain and Ireland (Preston *et al.* 2002).

Since 1988, NCC and its successor organisations (Countryside Council for Wales (CCW), English Nature (EN) and Scottish Natural Heritage (SNH)) have commissioned a substantial number of additional lake surveys, leading to the establishment of a much larger dataset. In particular, the Scottish Loch Survey, completed by SNH, extended coverage throughout Scotland, while CCW concentrated on collecting data from a representative series of sites in Wales (Allott & Monteith 1999; Allott *et al.* 2001). Surveys in England

focused on the areas with the highest concentrations of lakes and where management information was required. A comprehensive dataset has also been established for lakes in Northern Ireland (Wolfe-Murphy *et al.* 1992) but it was not included in this analysis.

In 1992 the European Community adopted *Council Directive 92/43/EEC on the conservation of natural habitats and of wild fauna and flora,* commonly known as the Habitats Directive. The main aim of this Directive is "to contribute towards ensuring biodiversity through the conservation of natural habitats and of wild fauna and flora in the European territory of the Member States to which the Treaty applies". It also brought an obligation for each member state to select, designate and protect a series of sites, to be called Special Areas of Conservation (SACs). In Britain, the first step in this process, site selection, involved making links between the JNCC lake classification scheme and the lake habitat types listed in the Directive, which were considered to be in need of conservation at a European level. Guidance is provided in this report on the relationship between the British lake groups described and Habitats Directive Annex I standing water habitats - Oligotrophic waters containing very few minerals of sandy plains (*Littorelletalia uniflorae*); Oligotrophic to mesotrophic standing waters with vegetation of the *Littorelletea uniflorae* and/or of the *Isoëto-Nanojuncetea*; Hard oligo-mesotrophic waters with benthic vegetation of *Chara* spp.; Natural eutrophic lakes with *Magnopotamion* or *Hydrocharition*-type vegetation; Natural dystrophic lakes and ponds; and coastal lagoons.

More recently, the Directive of the European Parliament and of the Council establishing a framework for Community action in the field of water policy (commonly known as the Water Framework Directive (WFD)) was adopted in 2003 (Pollard & Huxham 1998; Foster *et al.* 2001). Macrophytes and phytobenthos are included as biological elements on which assessments of ecological status will be based. Measurements of taxonomic composition and abundance are required as a basis for the ecological assessments. The JNCC therefore considered it timely for the production of a revised lake classification scheme based on its extended data holdings. As part of the preparation for the technical implementation of the WFD, close collaboration was developed between the conservation agencies and the regulatory agencies with regard to the characterisation and assessment of lakes as required by the Directive. This gave access to additional site data and opened a number of avenues for potential application of the classification scheme. Palmer (2001) highlighted a number of tasks that would make the JNCC lake classification approach more useful in the context of the WFD. These included a re-analysis of the enlarged dataset and the incorporation of macro-algae records at species level.

This report presents a description of the results of the new classification scheme based on the enlarged dataset. A revised key for the classification of newly surveyed sites has been produced. Individual ecological descriptions are given of the 11 distinct lake groups. Differences from the previous classification are highlighted. The earlier lake classification (Palmer 1992, Palmer *et al.* 1992) was used to derive Trophic Ranking Scores (TRSs) both for species and for sites. In this report, the TRSs have been replaced by a new scoring scheme, the Plant Lake Ecotype Index (PLEX), which reflects the complex response of freshwater plant assemblages to a large number of environmental variables, especially alkalinity and pH. Two case studies examining changes in PLEX values for lakes impacted by eutrophication and acidification are presented. The group composition and regional distribution of the surveyed lakes is considered. Finally, there is discussion of possible applications of the data collected and the resultant classification, in the context of the Habitats Directive, the Water Framework Directive and other conservation initiatives. Recommendations are made for future research and development.

The preliminary results of this research were presented at the Annual Scientific Meeting of the Freshwater Biological Association 5-6 September 2002, the symposium on "Typology and ecological classification of lakes and rivers" in Helsinki, 24-26 October, 2002 (Duigan & Phillips 2002) and the winter meeting of the British Ecological Society, 18-20 December, 2002 (Duigan *et al.* 2002). A summary publication for the scientific literature will be published in *Aquatic Conservation: Marine and Freshwater Ecosystems*.

Chapter 2: Survey method

The standing waters on which this revised classification is based were surveyed between 1975 and 2001 and included natural lakes, reservoirs, ponds, pools and gravel pits. Canals were included in the first classification scheme (Palmer 1992, Palmer et al. 1992), but it was decided to exclude them from the present analysis because the vegetation composition is likely to be influenced by an additional disturbance factor (boat traffic) (Willby *et al.* 2001). It is certain that a significant proportion of the lakes included will have been influenced by anthropogenic factors, such as eutrophication and atmospheric acid deposition. No attempt was made to identify sites that could be considered at "reference condition" in the context of the WFD. A standardised survey method for lake macrophytes was developed by the Nature Conservancy Council, and has subsequently been adopted by its successor organisations (e.g. Lassière 1998). It has also been used and modified by other organisations, as a means of lake characterisation and monitoring (e.g. Wolfe-Murphy *et al.* 1991; Parr *et al.* 1999). For the purpose of assessing ecological status, the European Committee for Standardization is currently developing a guidance standard for the surveying of macrophytes in lakes (CEN 2006) and this standard incorporates and develops the survey methods described here.

In general, the surveys were carried out between May and mid-September. Sites were surveyed from the lake shore and/or from a boat. The number of surveyors used was dictated by the size of the lake, the exact methods used, and health and safety regulations. The shore-based survey involved walking around the edge of the lake, between the upper limit of the inundation zone and maximum wading depth (circa 0.6m) and recording macrophyte distribution by eye. A bathyscope or other underwater viewing device was used, if available. Deeper water was sampled by means of a double headed rake (or grapnel) attached to a length of rope, and thrown from the lake shore into deeper water at regular intervals. Lakes with rocky substrates precluded the regular use of the rake because of the likelihood of losing the equipment. At some sites the extent of each macrophyte community will have been mapped. A series of target notes was usually compiled to describe the range and abundance of species at particular points around the lake and the location of rare species.

The boat-based survey techniques varied according to the shape and size of the lake and the weather conditions. If possible, attempts were made to cover the entire water area to record all the species present. An Eckman grab was sometimes used to sample macrophytes in deeper water alongside the boat. Following a complete shoreline walk, supplemented on occasions by a boat survey, the abundance of aquatic species recorded at the site was generally estimated on a semi-quantitative DAFOR scale where: D = Dominant; A = Abundant; F = Frequent; O = Occasional; R = Rare. A standardised recording sheet, including a plant species list, was developed by NCC, and subsequently modified to record additional site information at a country level. This allows information to be recorded on location, access, land use in catchment, uses and impacts on lake (e.g. abstraction, fishing, sewage discharge), substrate variability, and any notable bird, mammal, amphibian or dragonfly sighting made in the course of the survey. For the purposes of recording at most lakes, the plant list will have been divided into broad groups based on growth form and occurrence on the lake margins or offshore.

For the majority of the sites single measurements of pH and electrical conductivity were made in the field using a variety of hand-held meters. Alkalinity, nutrient and various other water chemistry parameters were measured at a smaller subset of sites. For the Scottish sites, these measurements were made from a single sample collected during the summer period. The majority of the water chemistry data for England are also single measurements. The water chemistry data used for Welsh sites represents the mean of four seasonal water measurements. The individual lake survey sheets are held by the relevant country conservation agency.

Chapter 3: The dataset, data analysis and the Plant Lake Ecotype Index (PLEX)

The dataset

All the survey data collected are stored in the JNCC GB Standing Waters Database. It was decided to include in the re-analysis only submerged and floating taxa, as emergent vegetation is subject to influences different from those experienced in open water (e.g. stock grazing) and it may not be a true reflection of conditions within the lake (Palmer *et al.* 1992). Careful consideration was given to where the taxa are usually found in the water, and reference was made to Preston & Croft (1997) in order to distinguish the submerged and floating taxa from emergent species. Querying the database and manipulating the output produced an Excel spreadsheet for re-analysis. The plant taxa include representatives of the angiosperms, bryophytes, hepatophytes and macroalgae (Appendix A). The taxonomic nomenclature follows Stace (1991), Preston (1995) and Paton (1999). Where more than one survey for a lake has been undertaken, the most recent survey data were used. It is acknowledged that some of the survey data are relatively old but the information is still a snapshot of a lake environment with a concurrent plant assemblage. However, the data will need to be updated for particular regions if contemporary comparisons between sites are required. Attempts were made to ensure that the taxon list was consistent with those used in the earlier classification (Appendix A).

Data analysis

The classification scheme is based on the results of an analysis of the species data by Two-Way Indicator Species Analysis, using the program TWINSPAN (Hill 1979). The analysis was done with a version of the program containing the corrections described in Oksanen & Minchin (1997).

The analysis was performed on data from 3447 lakes (310 England, 38 Wales and 3099 Scotland; see Figure 3.1). The species data for most sites were recorded on the DAFOR scale (see chapter 2). Some lakes had species just recorded as 'present'. These were stored in the data file as the value 1.5 (half way between 1 for Rare and 2 for Occasional). The pseudospecies cut levels (1, 2, 3, 4 and 5) for the TWINSPAN analysis were chosen to match the DAFOR scale; those recorded as 1.5 would thus be included in the Occasional pseudospecies.

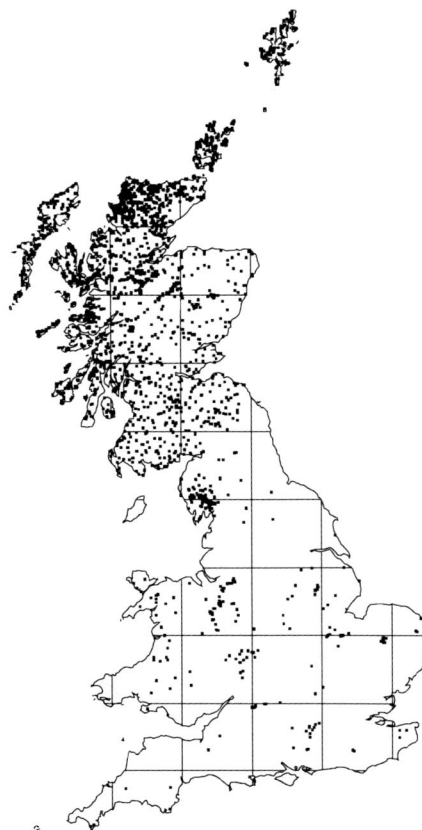

Figure 3.1. Distribution of lakes included in the analysis. *The gridlines delineate 100km squares.*

The results of the analysis are presented graphically in Figure 3.2 as a dendrogram. End points of the dendrogram were chosen as recognisable lake types, based on survey experience. The groups either comprised fewer than 250 sites or were formed by the fifth division of the TWINSPAN analysis. These end points were then designated as the major lake groups: A-J. Group C, which contains 45% of the sites, was further divided into two subgroups, C1 and C2, based on the sixth division of the TWINSPAN analysis.

10

Figure dendrogram labels:

n=3447

J. bulbosus 1 — n=3412
R. maritima 1 / P. pectinatus 1 — n=35

J. bulbosus 1
P. polygonifolius 1
L. uniflora 1
M. alterniflorum 1
L. dortmanna 1
S. angustifolium 1 — n=2779

L. minor 1 — n=633

Chara sp. 1
M. spicatum 1
P. filiformis 1
P. pectinatus 1
P. pusillus 1 — n=203

J

L. uniflora 1
M. alterniflorum 1
P. natans 1
L. dortmanna 1
P. polygonifolius 1 — n=2557

Sphagnum sp. 1 — n=222

A

L. minor 1 — n=430

P. amphibia 1
E. canadensis 1
L. minor 1
N. lutea 1
P. natans 1 — n=329

I

L. dortmanna 1
J. bulbosus 1
P. polygonifolius 2 — n=2001

Nitella sp. 1
C. hamulata 1
F. antipyretica 1
G. fluitans 1 — n=556

G. fluitans 1
C. stagnalis 1 — n=101

H

L. uniflora 1
M. alterniflorum 1
L. dortmanna 2
I. lacustris 1
S. angustifolium 1

Sphagnum sp. 1 — n=426

B

Nitella 1
C. hamulata 1 — n=370

Chara sp. 1
P. perfoliatus 1
P. filiformis 1
P. gramineus 1 — n=186

D   E

N. lutea 1
N. alba 2 — n=48

P. natans 1
G. fluitans 1
E. canadensis 1
P. crispus 1
P. obtusifolius 1 — n=281

F   G

n=1575

P. natans 1
M. alterniflorum 1
P. polygonifolius 1
L. dortmanna 1
L. uniflora 2
N. alba 1

Sphagnum sp. 1 — n=256

n=1319

C1   C2

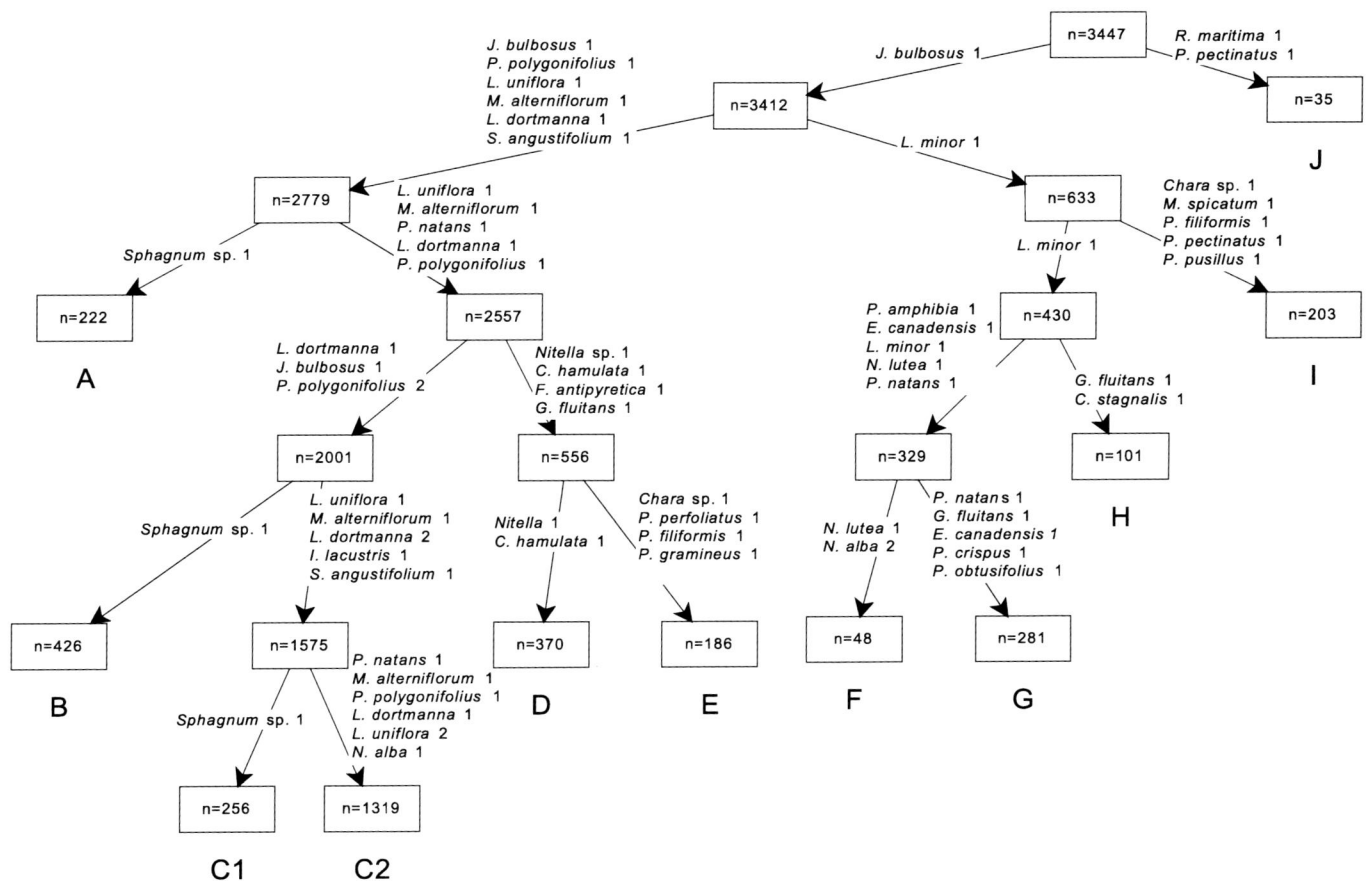

Figure 3.2. Dendrogram of TWINSPAN analysis showing major groups A-J
The differentiating plant taxa are shown on the branches.

Box plots were produced for each environmental variable, with the sites arranged by lake group (Figures 4.1-4.5). These plots are comprised of a two-section box for each group indicating the second and third quartiles of the data (the Interquartile Range, or IQR), separated by a line showing the median value. Two T-shaped lines extending above and below the box indicate the extent of the adjacent values: these are defined as the highest or lowest observation that lies within the inner fences. The inner fences themselves are those boundaries that are placed 1.5 times the IQR above and below the box itself. Outliers and extreme outliers are indicated by filled and empty dots.

Two Canonical Correspondence Analyses (CCA: ter Braak 1986) were performed, using the program MVSP (Kovach 2001). The CCA analyses were limited to those sites that had data for all variables, and brackish water sites (Group J) were excluded; this reduced the data set to 1035 sites and 84 taxa. The first CCA was done using the environmental variables conductivity, alkalinity and lake surface area (all log transformed to reduce skewness) as well as pH and altitude.

In order to incorporate geographical data into the analysis another CCA was performed with the addition of categorical variables for each 100km square of the UK National Grid Reference scheme. Then, for each site, a value of 1 was placed in the categorical variable corresponding to its grid reference (e.g. variable SH for a site with grid reference SH646595); all other grid square variables contained 0 for the site.

Chi square tests were also calculated to determine whether the distribution of each species between the five ecotype categories deviated from a uniform distribution. These tests were performed on all species that occurred in more than 25 non-brackish sites. All tests gave significant results, most with extremely low probabilities, indicating that all species had differential distribution among the ecotypes.

Plant Lake Ecotype Index

Plant Lake Ecotype Index (PLEX) scores for each species were calculated using a modification of the Trophic Ranking Scores (TRS) method described by Palmer *et al.* (1992). However, instead of ascribing the lakes groups to trophic classes, they have been grouped into distinct ecotype categories. This new scheme is presented as an index of lake environments based on macrophyte composition.

First, each TWINSPAN end group was assigned to one of five ecotype categories: dystrophic lakes with low plant diversity (Group A); heathland-associated soft waters in the lowlands and mountains (Groups B & C); circumneutral, mid to low altitude lakes with a diverse assemblage of plants (Groups D and E); hardwater, lowland lakes with low to moderate plant diversity (Groups F, G and H); and hardwater, lowland lakes with *Chara* (Group I). Brackish water sites (Group J) were excluded, as in Palmer *et al.* (1992). Any species with fewer than 25 non-brackish occurrences was also excluded.

Next, for each species the expected number of occurrences in each ecotype category, assuming uniform distribution across the categories, was calculated. The ratio of observed versus expected numbers was then calculated for each ecotype category. If the ratio was greater than 2.0 (i.e. the species occurred twice as often as expected), this was counted as a strong association between the species and the ecotype category. A ratio of between 1.1 and 2.0 was considered to be a weak association.

For further manipulation, the ecotype categories were then assigned code-letters (taken from the end of the alphabet, to avoid confusion with the TWINSPAN end groups), as follows:

V - Dystrophic lakes, with low plant diversity;
W - Heathland-associated soft waters in the lowlands and mountains, with low plant diversity;
X - Mid to low altitude lakes, with a diverse assemblage of plants;
Y - Hardwater, lowland lakes, with low to moderate plant diversity;
Z - Hardwater, lowland lakes with *Chara*.

Each species was then assigned one or more of these letters to indicate their ecotype preferences. For strong associations an upper-case capital letter corresponding to the ecotype category (V, W, X, Y, or Z) was assigned to that species. If the association was weak, lower-case letters were assigned. So, for example, *Sphagnum* spp., which has a strong association with peatland lakes (3.40 observed/expected ratio) and a weak association with soft waters (1.18 observed/expected ratio), can be assigned the ecotype code **Vw**. *Nuphar lutea*, which has a weak association with X (mid to low altitude lakes, with a diverse assemblage of plants; 1.60 observed/expected ratio) and a strong association with Y (hardwater, lowland lakes, with low to moderate plant diversity; 3.32 observed/expected ratio), has a code of **xY**.

These ecotype codes were then converted to PLEX values by first assigning each letter a value according to the following table:

Ecotype code	PLEX Value
V	1
v	2
w (adjacent to V or v)	3
W (or w with no V, v, X or x)	4
w (adjacent to X or x)	5
x (adjacent to W or w)	6
X (or x with no W, w, Y or y)	7
x (adjacent to Y or y)	8
y (adjacent to X or x)	9
Y (or y with no X, x, Z or z)	10
y (adjacent to Z or z)	11
z	12
Z	13

The numbers corresponding to the ecotype codes for each species were then summed and the mean was calculated. So, for example, for a species with the code **VWx** the PLEX would be $(1 + 4 + 6)/3 = 3.7$.

Note that all possible values from the above table must be applied. For a code of **wxY** the **x** can be scored as both 6 and 8, so both are used. The resulting PLEX score is $(5 + 6 + 8 + 10) / 4 = 7.25$.

Finally, to provide an index on a scale from 1-10 (rather than the more unusual 1-13 for the raw scores) the above scores are rescaled to 10 by dividing by 13 and multiplying by 10. The final PLEX scores for the species are listed in the table in Annex B.

12

Once the PLEX score has been calculated for each species an average PLEX score for a site can be calculated from the assemblage of plants. In the following example the PLEX score of a hypothetical site with five species is calculated:

Species	Ecotype code	Raw PLEX score	PLEX 1-10
Potamogeton perfoliatus	XZ	7+13 / 2 = 10	(10/13)*10 = 7.69
Callitriche hamulata	Xy	7+9 / 2 = 8	(8/13)*10 = 6.15
Littorella uniflora	wx	5+6 / 2 = 5.5	(5.5/13)*10 = 4.23
Isoetes lacustris	wx	5+6 / 2 = 5.5	(5.5/13)*10 = 4.23
Juncus bulbosus	w	4	(4/13)*10 = 3.08
Average PLEX			**5.08**

The box plots in Figure 3.3 show that the different lake groups have differences in median PLEX scores, which indicates the reliability of PLEX.

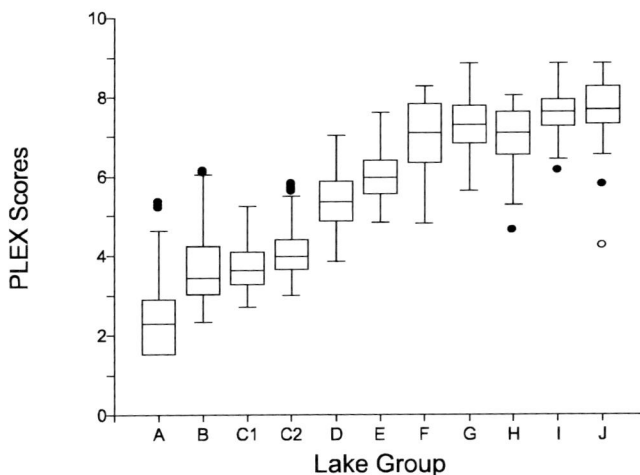

Figure 3.4. PLEX scores plotted against pH.
*Regression equation: PLEX = -1.53 + 0.96*pH, correlation - 0.707.*

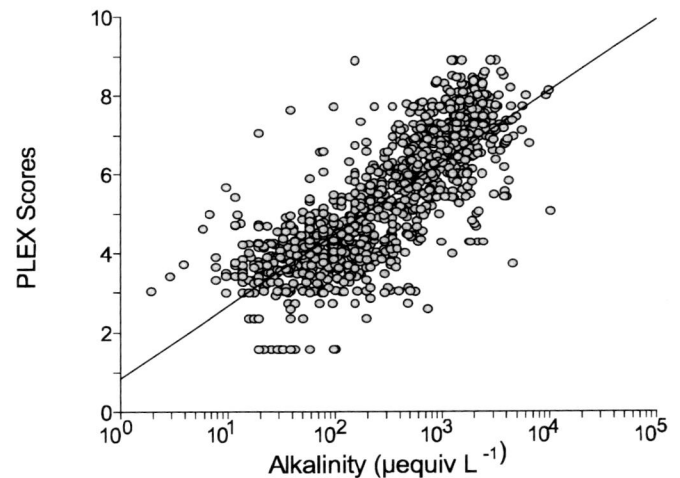

Figure 3.3. Ranges of PLEX scores for lake groups.

Changes in the PLEX value for a particular site will indicate environmental change meriting further investigation. In this way, it is comparable to the earlier TRS scheme which was advocated as a simple 'early warning' system, or as one element in a multimetric approach to monitoring water quality (Palmer 2001). In addition, there is evidence that PLEX is an indicator of base status (alkalinity and pH) as shown in Figures 3.4 and 3.5, and is likely be co-correlated with nutrient and/or acid status. In Chapter 6 changes in PLEX scores developed from monitoring data for two lakes impacted by enrichment and acidification are discussed.

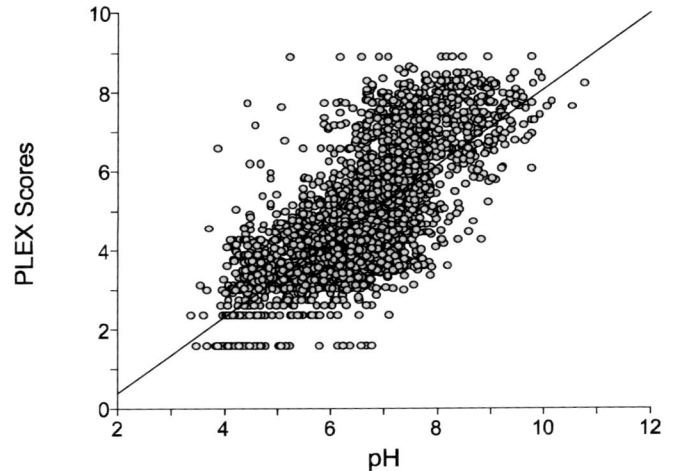

Figure 3.5. PLEX scores plotted against alkalinity.
*Regression equation: PLEX = 0.84+ 0.78*Ln(Alkalinity), correlation -0.783.*

Chapter 4: Revised classification system and lake key

The TWINSPAN results were used to identify 10 major sub-groups, with the large number of lakes in Group C divided into C1 and C2 on the basis of taxon richness (Figure 3.2; see above).

Table 4.1 shows the submerged and floating plant taxa occurring at a constancy of more than 20% in the TWINSPAN end groups chosen. There is an obvious visual progression of species in terms of frequency of occurrence across the groups and down the list of plant taxa.

A classification key was derived with reference to the divisions in the dendrogram and the most frequently occurring plant taxa in each group (Table 4.2).

Figures 4.1-4.6 illustrate the ranges of altitude (m), surface area (ha), pH, conductivity (μS cm^{-1}), alkalinity (μequiv L^{-1}) and taxon richness (= total number of plant taxa listed in annex A). The highest median values for altitude are found in Group A and Group C1 which reflects the large number of mountain lakes assigned to these groups (Figure 4.1). Median altitude values decrease almost progressively across Groups C2 to J. The lowest altitude values are associated with groups dominated by sites in lowland or coastal locations (Group I and Group J). The lakes with the greatest surface area occur in Groups E, D and I, while the smallest lakes are found in Group A (Figure 4.2). The latter association is consistent with the ecology of these small lakes, as described in Chapter 5. As expected, there appears to be an association between the pH and conductivity values of the individual lake groups (Figures 4.3 and 4.4). Group A has the lowest median values, while Group J has the highest median values for these two water chemistry parameters. This is turn links with the alkalinity values for each group shown in Figure 4.5, with Groups A and C1 having the lowest median alkalinity values, while the highest median values are found in Groups F, I and J. Finally, the median value for taxon richness is highest for the relatively large lakes in Group E (Figure 4.6). Groups A and J are species poor.

Figure 4.7 summarises the number of sites per lake group in Scotland, England and Wales. Most of the Group A lakes occur in Scotland. Groups B and D are represented in all three countries. Groups C1 and C2 are among the most common lake groups in all three countries, with C1 always occurring with less frequency than C2. Group E is a relatively rare lake group in all three countries. In contrast with Scotland and Wales,

lake groups F and G are dominant in England. Group H representatives are almost exclusive to Scotland while Group I lakes occur in all three countries. Group J lakes are relatively rare but best represented in Scotland. Representatives of groups A, H and J are absent from Wales.

The results of the CCA of the main environmental variables are shown in Figure 4.8 and Table 4.3. The amount of total species variability accounted for by the five constrained axes of the first analysis was just 7.63%. When the geographical data are included, through the use of variables for each 100km grid reference square, the percentage increased to 11.66% (Table 4.4). Although low values are not unusual when there are large numbers of taxa, in this case the amount of variation is just slightly larger than if the total variation were spread evenly across all 84 taxa/axes (5.95% for five axes). This indicates that there are many other factors affecting species distribution than just the environmental variables used here.

However, the CCA results do show correspondence to the lake groups defined in this report. In Figure 4.8 the points for each group are distinguished by different shapes and shading. The points for each group are also encircled to show their distribution more clearly. This demonstrates a progression from right to left through the groups in the order A-I, with Group A to the furthest right and Group I to the left.

The first axis corresponds to a gradient from high altitude and low alkalinity, pH and conductivity on the right, to lower altitudes and higher chemical parameters on the left. These trends and their correspondence to the groups can also be seen in the box plots in Figures 4.1 and 4.3-4.5. The second axis represents a gradient from low surface area at the top to high surface area at the bottom. The sites from groups A and B lie primarily or entirely on the upper half of this gradient, showing a correlation with smaller lakes, while Group E lies primarily to the lower end, correlated with larger lakes.

The results of the CCA including the grid reference data were less satisfactory. The first axis showed a similar trend to the CCA described above, but after that each axis had moderate correlations with particular grid reference squares (see Table 4.4). For example, Axis 2 was dominated by a single site (Kenfig Pool) that was the only site included in the CCA that occurred in grid square SS, while Axis 3 had correlations with the squares covering north and west Wales and Cheshire.

14

Table 4.1. Constancy table for site groups
Submerged and floating plants constancy classes over 20% only.

Taxon	A	B	C1	C2	D	E	F	G	H	I	J
Sphagnum (aquatic indet.)	V	IV	III	II							
Juncus bulbosus	III	IV	V	V	III	III					
Potamogeton polygonifolius		IV	II	IV	II	II					
Potamogeton natans		III		IV	III	III		III		II	
Nymphaea alba		III		II	II		III				
Eleogiton fluitans		II		II							
Utricularia minor		II									
Littorella uniflora			III	V	IV	V				II	
Myriophyllum alterniflorum			II	IV	IV	V					
Glyceria fluitans				II	IV	III		III	IV	II	
Fontinalis antipyretica				II	IV	III					
Callitriche hamulata					IV			II			
Sparganium angustifolium			IV	III	III	II					
Nitella spp.			II	III	II						
Lobelia dortmanna			II	V	II						
Isoetes lacustris			II	III	II	II					
Elodea canadensis					II		II	III		II	
Callitriche stagnalis					II	II	III	II	IV	II	II
Chara spp.					II	IV		II		IV	
Potamogeton berchtoldii					II	II		II		II	
Potamogeton perfoliatus					II	IV				II	
Subularia aquatica				II							
Lemna minor							IV	IV		II	
Persicaria amphibia					II	III		III		II	
Myriophyllum spicatum								II		III	
Potamogeton crispus								II		II	
Lemna trisulca								II			
Potamogeton obtusifolius								II			
Nuphar lutea							V				
Zannichellia palustris							II			II	
Potamogeton filiformis						III				II	
Potamogeton gramineus						III					
Callitriche hermaphroditica					II					II	
Apium inundatum					II						
Potamogeton gramineus x perfoliatus							II				
Potamogeton pectinatus										III	III
Enteromorpha spp.											III
Ruppia maritima											III
Fucoid algae											II
Potamogeton pusillus										III	
Ranunculus baudotii										II	

Constancy classes: V - >80 to 100%; IV - >60 to 80%; III - >40 to 60%; II - >20 to 40%.

Table 4.2. Key to standing water site types using submerged and floating taxa.
The key is used to classify lakes using macrophyte data collected using the standard method described in Chapter 2. Presence/absence records are used unless an indication of minimum abundance levels is given, according to the DAFOR scale. Score -1 for every record of a negative indicator; score +1 for every record of a positive indicator.

	Negative indicators (-1)	Positive indicators (+1)	Score	Go to	Group
1	*Juncus bulbosus*	*Potamogeton pectinatus*	1 or less	2	-
		Ruppia maritima	2 or more	-	J
2	*Juncus bulbosus*	*Lemna minor*	-1 or less	3	-
	Lobelia dortmanna		0 or more	4	-
	Littorella uniflora				
	Myriophyllum alterniflorum				
	Potamogeton polygonifolius				
	Sparganium angustifolium				
3	*Sphagnum* spp.	*Littorella uniflora*	-1	-	A
		Lobelia dortmanna	0 or more	5	-
		Myriophyllum alterniflorum			
		Potamogeton natans			
		Potamogeton polygonifolius			
4	*Lemna minor*	*Chara* spp.	1 or less	6	-
		Myriophyllum spicatum	2 or more	-	I
		Potamogeton filiformis			
		Potamogeton pectinatus			
		Potamogeton pusillus			
5	*Juncus bulbosus*	*Callitriche hamulata*	0 or less	7	-
	Lobelia dortmanna	*Fontinalis antipyretica*	1 or more	8	-
	Potamogeton polygonifolius (at least Occasional)	*Glyceria fluitans*			
		Nitella spp.			
6	*Elodea canadensis*	*Callitriche stagnalis*	0 or less	9	-
	Lemna minor	*Glyceria fluitans*	1 or more	-	H
	Nuphar lutea				
	Persicaria amphibia				
	Potamogeton natans				
7	*Sphagnum* spp.	*Isoetes lacustris*	0 or less	-	B
		Littorella uniflora	1 or more	10	C
		Lobelia dortmanna			
		Myriophyllum alterniflorum			
		Sparganium angustifolium			
8	*Callitriche hamulata*	*Chara* spp.	0 or less	-	D
	Nitella spp.	*Potamogeton filiformis*	1 or more	-	E
		Potamogeton gramineus			
		Potamogeton perfoliatus			
9	*Nuphar lutea*	*Glyceria fluitans*	-1 or less	-	F
	Nymphaea alba (at least Occasional)	*Elodea canadensis*	0 or more	-	G
		Potamogeton crispus			
		Potamogeton natans			
		Potamogeton obtusifolius			
10	*Sphagnum* spp.	*Littorella uniflora* (at least Occasional)	0 or less	-	C1
		Lobelia dortmanna (at least Occasional)	1 or more	-	C2
		Myriophyllum alterniflorum			
		Potamogeton natans			
		Potamogeton polygonifolius			
		Nymphaea alba			

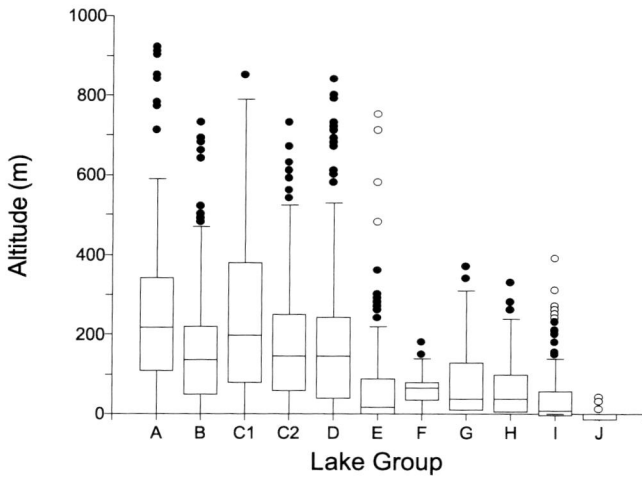

Figure 4.1. Ranges of altitude for lake groups.

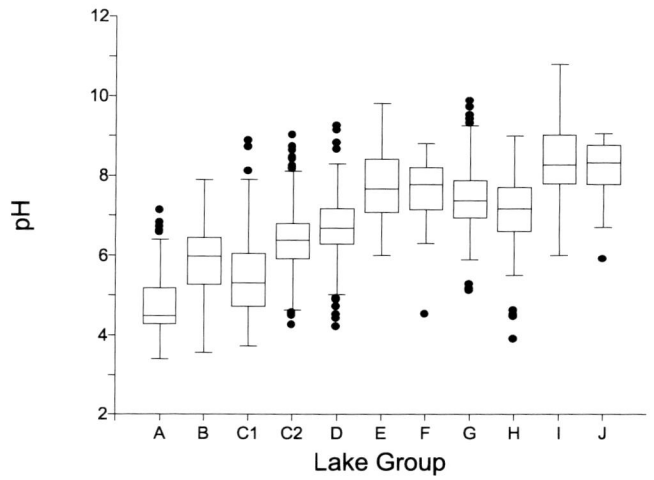

Figure 4.3. Ranges of pH for lake groups.

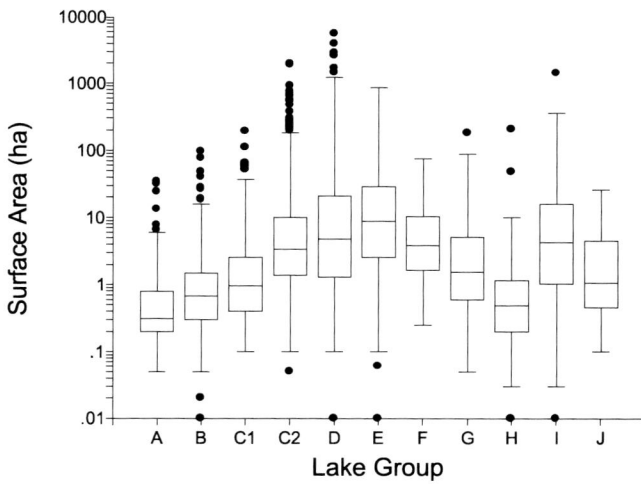

Figure 4.2. Ranges of surface area for lake groups.

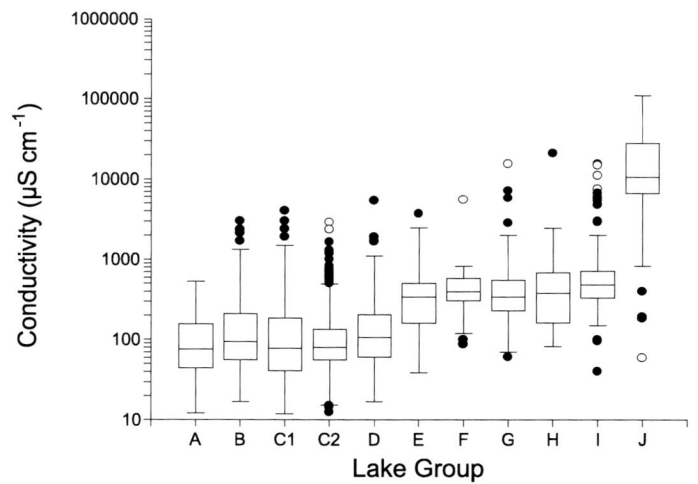

Figure 4.4. Ranges of conductivity for lake groups.

17

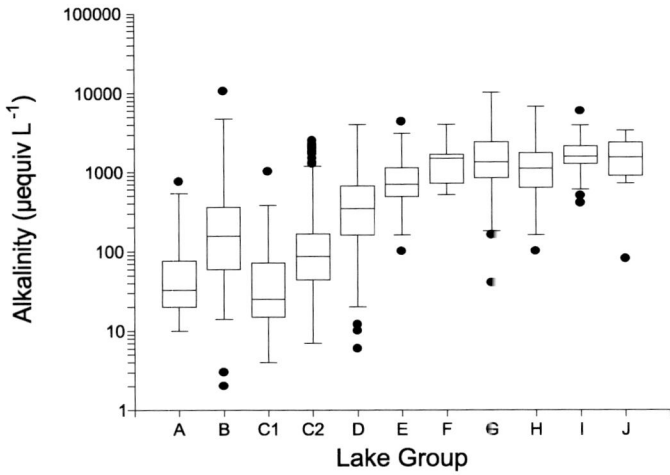

Figure 4.5. Ranges of alkalinity for lake groups.

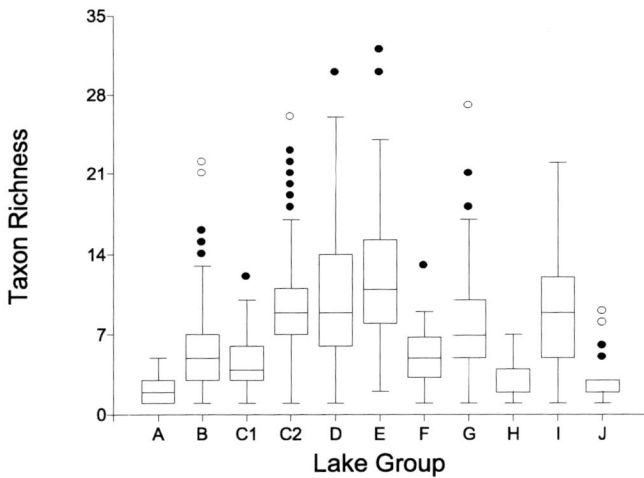

Figure 4.6. Ranges of taxon richness for lake groups.

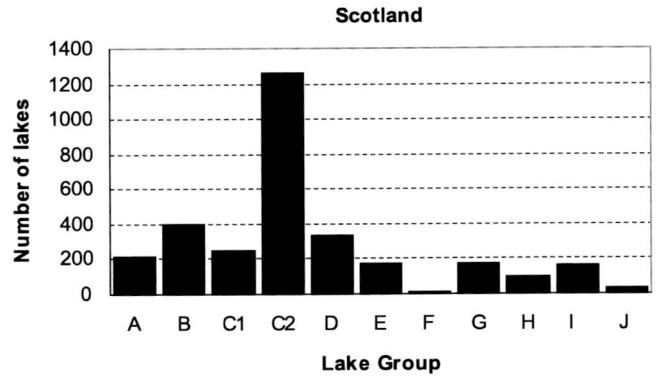

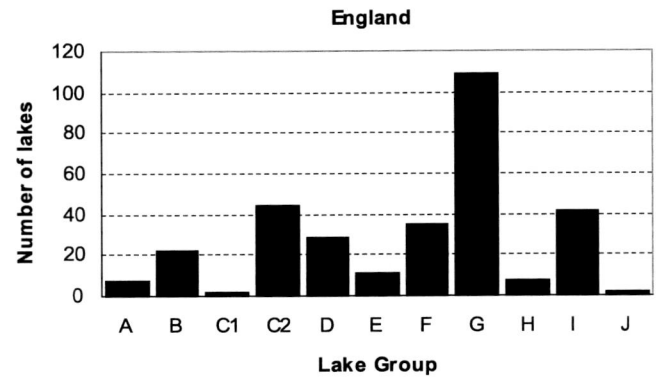

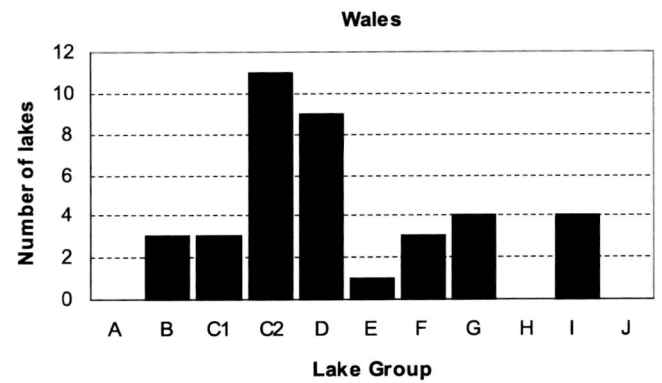

Figure 4.7. Number of sites per lake group for Scotland, England and Wales.

18

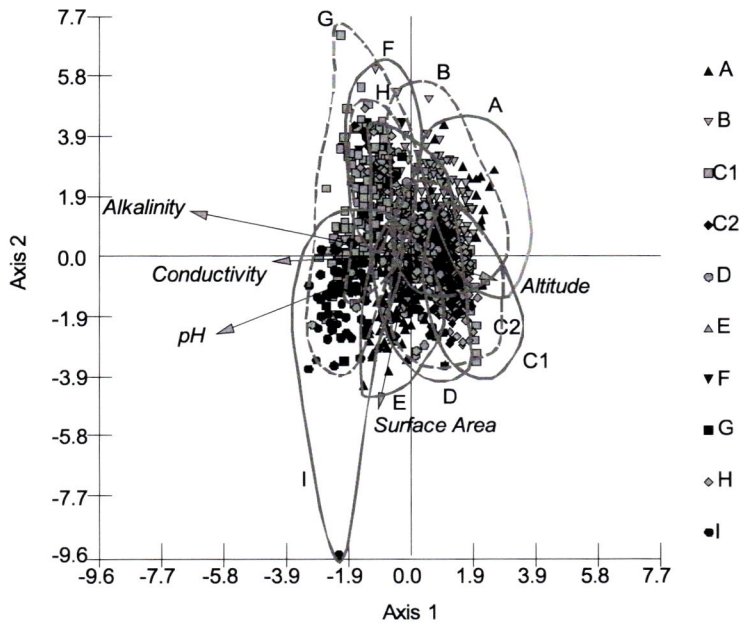

Figure 4.8. Ordination diagram of a CCA analysis,
including the major environmental variables

Table 4.3. Statistics for the CCA analysis in Figure 4.8					
	Axis 1	*Axis 2*	*Axis 3*	*Axis 4*	*Axis 5*
Eigenvalues	0.395	0.097	0.049	0.038	0.018
Percentage	5.058	1.237	0.627	0.483	0.227
Cumulative Percentage	5.058	6.295	6.922	7.405	7.632
Cumulative Constrained Percentage	66.280	82.489	90.707	97.030	100
Species-Environment Correlations	0.858	0.639	0.490	0.487	0.372
Intraset correlations					
Altitude	0.379	-0.108	0.535	0.711	-0.231
Surface Area	-0.143	-0.703	0.488	-0.351	0.353
Conductivity	-0.608	-0.025	-0.661	0.104	0.427
pH	-0.857	-0.358	-0.161	-0.007	-0.334
Alkalinity	-0.975	0.204	0.049	0.061	0.035

Table 4.4. Statistics for the CCA analysis incorporating 100km grid squares

	Axis 1	Axis 2	Axis 3	Axis 4	Axis 5	Axis 6	Axis 7	Axis 8	Axis 9	Axis 10
Eigenvalues	0.455	0.160	0.142	0.085	0.067	0.066	0.050	0.041	0.035	0.031
Percentage	5.835	2.056	1.824	1.090	0.859	0.840	0.644	0.519	0.447	0.392
Cumulative Percentage	5.835	7.892	9.715	10.805	11.664	12.504	13.148	13.667	14.114	14.506
Cumulative Constrained Percentage	33.762	45.661	56.213	62.521	67.489	72.347	76.074	79.077	81.664	83.933
Species-Environment Correlations	0.906	0.902	0.756	0.669	0.612	0.551	0.524	0.524	0.529	0.550
Intraset correlations										
Altitude	-0.328	0.003	0.124	0.368	-0.491	-0.186	-0.160	0.257	0.019	0.016
Surface Area	0.112	0.154	-0.441	0.480	0.207	-0.372	-0.171	-0.073	0.156	0.106
Conductivity	0.538	-0.007	-0.299	-0.414	0.032	0.276	0.206	-0.012	0.034	-0.015
pH	0.769	0.060	-0.462	-0.036	0.111	0.116	-0.117	-0.089	-0.064	-0.092
Alkalinity	0.902	-0.025	-0.081	-0.245	0.142	0.015	0.010	0.082	-0.003	-0.017
HP	0.022	0.045	-0.354	-0.011	-0.341	0.320	0.456	-0.282	0.141	0.098
HT	0.015	0.002	-0.128	-0.039	0.023	-0.090	0.092	0.087	0.006	0.058
HU	0.005	0.021	-0.425	-0.138	-0.024	-0.202	0.265	0.373	-0.106	0.105
HZ	-0.009	-0.003	-0.002	-0.027	-0.013	0.001	0.034	0.011	0.005	-0.009
NA	-0.049	0.011	-0.046	-0.007	-0.031	0.045	0.043	-0.070	0.002	0.039
NB	-0.154	0.003	-0.055	-0.053	0.061	-0.072	-0.032	0.014	-0.026	0.006
NF	-0.027	-0.004	-0.105	-0.142	0.178	0.034	-0.013	-0.050	-0.037	-0.018
NG	-0.155	0.012	0.000	-0.056	0.013	0.106	-0.063	-0.008	0.011	-0.004
NH	-0.092	-0.018	-0.032	-0.137	-0.141	-0.070	0.094	0.280	-0.094	-0.090
NJ	0.126	-0.057	0.073	-0.024	-0.127	-0.129	-0.038	0.071	-0.092	-0.064
NK	0.087	-0.037	0.064	0.035	-0.109	-0.045	-0.101	-0.052	-0.099	-0.097
NL	0.076	0.010	-0.206	-0.205	0.095	0.046	0.041	0.109	-0.132	-0.007
NM	-0.482	0.060	0.000	-0.075	0.221	0.364	-0.369	-0.223	0.132	-0.071
NN	-0.209	0.025	-0.006	0.170	-0.147	-0.088	-0.027	0.076	-0.187	-0.027
NO	0.236	0.025	-0.023	0.132	-0.109	0.141	-0.276	0.095	-0.116	-0.139
NR	-0.044	-0.001	-0.044	-0.283	0.138	-0.056	0.063	0.045	0.052	0.045
NS	0.323	-0.152	0.181	0.059	-0.014	-0.303	0.171	-0.266	0.370	-0.139
NT	0.408	-0.045	0.132	0.119	-0.383	0.118	-0.330	0.170	-0.143	0.149
NW	0.029	-0.014	0.019	-0.024	0.071	-0.021	0.006	-0.046	-0.057	0.041
NX	0.118	-0.114	0.256	0.001	0.098	-0.191	0.240	-0.198	0.182	0.120
NY	0.113	-0.048	0.103	0.042	-0.041	-0.080	-0.008	-0.041	0.107	-0.007
SH	0.055	0.166	0.005	0.423	0.134	0.092	0.194	-0.002	-0.234	-0.409
SJ	0.065	0.050	0.086	0.345	0.331	0.286	0.266	0.479	0.270	-0.223
SN	-0.055	0.024	0.096	0.331	0.143	-0.061	0.242	-0.157	-0.573	0.461
SO	0.103	0.119	0.047	0.136	0.107	0.252	-0.129	0.239	0.309	0.651
SS	0.080	0.945	0.173	-0.159	-0.070	-0.135	0.032	-0.060	0.031	-0.018

Chapter 5: Individual descriptions of the lake groups

In this chapter each lake group is described in terms of its distribution, environmental characteristics and characteristic plant taxa. Plant taxon constancy tables are produced for each group, with taxa having a constancy >20% indicated in bold. Links are made with the previous British lake classification (Palmer 1992, Palmer *et al*. 1992) and the National Vegetation Classification (NVC) (Rodwell *et al*. 1995). In addition, attempts are made to link the lake habitats listed in the EC Habitats Directive with the new British classification but, as expected of all ecological datasets, there is only partial correspondence between the classification schemes used. Examples of lakes that are candidate SACs under the Habitats Directive are given. Details are also given of the occurrence in each lake group of submerged and floating species that are internationally or nationally protected, categorised as Red List, Nationally Rare or Nationally Scarce, or are plants for which Britain has 'Special Responsibility'.

Table 5.1 gives the rarity status of native submerged and floating standing water plant species, many of which were included in the analysis. The information is taken from *LACON: Lake Assessment for Conservation* (Palmer, in prep.). Hybrids and mosses are not included. Data for vascular plants are based on the *New Atlas of the British and Irish Flora* (Preston *et al*. 2002). Species in the first five rarity categories are listed on the JNCC web site: www.jncc.gov.uk. The constancy of species in Table 5.1 within each lake group is given in Annex B.

Table 5.1. Rarity status of submerged and floating standing water plants

		EC/ Bern	WCA Sch	Red List	NR/ DD	Nat Sc	Sp Res	BAP	Dist	Occ
VASCULAR PLANTS										
Alisma gramineum	Ribbon-leaved water-plantain		8	CR				SAP	E	1
Apium inundatum	Lesser marshwort						SR		E S W	186
Callitriche truncata	Short-leaved water-starwort					NS			E W	1
Elatine hydropiper	Eight-stamened waterwort					NS			E S W	15
Eleogiton fluitans	Floating club-rush						SR		E S W	523
Eriocaulon aquaticum	Pipewort				NR		SR		S	48
Hydrilla verticillata	Esthwaite waterweed				NR				Ex S	0
Luronium natans	Floating water-plantain	EC/B	8			NS	SR	SAP	E Si W	8
Najas flexilis	Slender naiad	EC/B	8			NS		SAP	S	22
Najas marina	Holly-leaved naiad		8	VU				SAP	E	0
Nuphar pumila	Least yellow water-lily					NS			E S	58
Nymphoides peltata	Fringed water-lily					NS			E Si Wi	11
Pilularia globulifera	Pillwort					NS		SAP	E S W	25
Potamogeton acutifolius	Sharp-leaved pondweed		VU						E	0
Potamogeton coloratus	Fen pondweed					NS			E S W	9
Potamogeton compressus	Grass-wrack pondweed					NS		SAP	E S W	1
Potamogeton epihydrus	American pondweed			VU			SR		Ei S	1
Potamogeton filiformis	Slender-leaved pondweed					NS			E S	169
Potamogeton friesii	Flat-stalked pondweed					NS			E S W	28
Potamogeton rutilus	Shetland pondweed				NR		SR	SAP	Ei S	1
Ranunculus hederaceus	Ivy-leaved crowfoot						SR		E S W	74
Ranunculus omiophyllus	Round-leaved crowfoot						SR		E S W	18
Ranunculus tripartitus	Three-lobed crowfoot			VU				SAP	E W	0
Ruppia cirrhosa (spiralis)	Spiral tasselweed					NS			E S	4
Stratiotes aloides	Water-soldier				NR				E Si Wi	0
Wolffia arrhiza	Rootless duckweed					NS			E W	1

21

Table 5.1. *(continued)*

		EC/ Bern	WCA Sch	Red List	NR/ DD	Nat Sc	Sp Res	BAP	Dist	Occ
LIVERWORTS										
Ricciocarpos natans	Fringed heartwort					NS			E	0
CHAROPHYTES										
Chara aculeolata (pedunculata)	Hedgehog stonewort					NS			E S W	1
Chara baltica	Baltic stonewort				VU			SAP	E S W?	0
Chara canescens	Bearded stonewort		8	EN				SAP	E S	0
Chara connivens	Convergent stonewort			EN				SAP	E	0
Chara curta	Lesser bearded stonewort					NS	SR	SAP	E S W	7
Chara fragifera	Strawberry stonewort			VU					E	0
Chara intermedia	Intermediate stonewort			EN					E	0
Chara muscosa	Mossy stonewort				DD		SR	SAP	S?	0
Chara rudis	Rugged stonewort				NR				E S Wx	2
Lamprothamnium papulosum	Foxtail stonewort		8		NR			SAP	E S	0
Nitella confervacea	Least stonewort				NR				S	7
Nitella flexilis	Smooth stonewort					NS			E S W	0
Nitella gracilis	Slender stonewort			VU				SAP	Ex S W	3
Nitella mucronata	Pointed stonewort					NS			E S W	1
Nitella tenuissima	Dwarf stonewort			EN				SAP	E W	0
Nitellopsis obtusa	Starry stonewort			VU				SAP	E	0
Tolypella glomerata	Clustered stonewort					NS			E S W	0
Tolypella intricata	Tassel stonewort			EN				SAP	E	0
Tolypella nidifica	Bird's-nest stonewort			EN				SAP	Ex S	0
Tolypella prolifera	Great tassel stonewort			EN				SAP	E	0

Explanatory notes to Table 5.1

EC/Bern
Species native to the UK and listed in Annexes IIb and IVb of the Habitats Directive and Appendix I of the Bern Convention.

WCA Sch
Species included in Schedule 8 of the Wildlife and Countryside Act 1981 and also protected under the Nature Conservation (Scotland) Act 2004.

Red List
Species included in British Red Lists.
Threat categories as in IUCN Species Survival Commission (2000) and Wigginton (1999): CR - Critically Endangered, EN - Endangered, VU - Vulnerable.

NR/DD
Nationally Rare (NR) species (previously called Near Threatened) are without an IUCN Red List designation in Britain, but have been recorded as native since 1986 in 15 or fewer 10 x 10 km squares in Britain.
Data Deficient (DD) species are those for which there is insufficient information to make an adequate assessment of threat.

Nat Sc
Nationally Scarce species.
These are plants that have been recorded as native since 1986 in 16 to 100 10 x 10 km squares in Britain and are without an IUCN Red List designation. Nationally Scarce vascular plants are listed in Stewart *et al.*, 1994, but a few designations have been changed as a result of recent data in Preston *et al.* (2002).

Spec Resp
Species for which Britain has Special Responsibility.
Species that are endemic or near-endemic to Europe and for which Britain has 'special responsibility' because it supports a high proportion (certainly or probably more than 25%) of the European population (Chris Preston and David Pearman, pers. com. for vascular plants; Stewart & Church, 1992, for charophytes).

BAP
Priority plant species with a current Species Action Plan under the UK Biodiversity Action Plan.

Dist
Species distribution: England (E), Scotland (S), Wales (W). Species extinct in a country are indicated by x. Species present only as introductions are indicated by i.

Occ
The number of occurrences in the dataset.

Group A: Small, predominantly northern dystrophic peat or heathland pools, dominated by *Sphagnum* spp.

No. of sites = 222

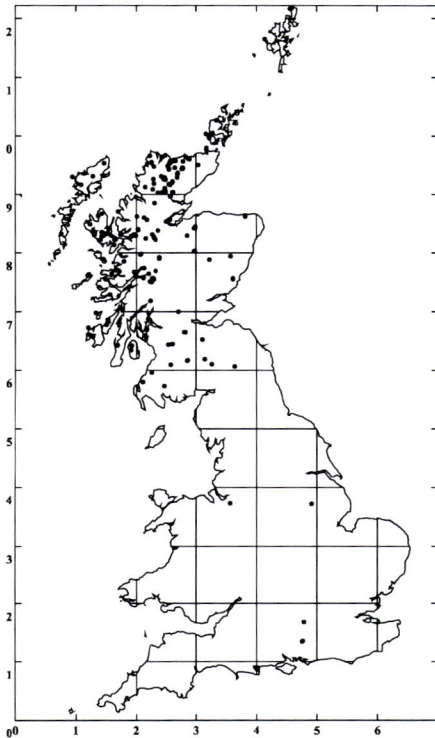

Summary features

Variable	n	Mean	Standard Deviation	Minimum	Maximum
Altitude	222	259.19	203.90	0.00	920.00
Surface area	220	1.13	3.65	0.05	34.00
pH	169	4.75	0.74	3.40	7.12
Conductivity (μS cm^{-1})	166	120.10	106.22	12.10	530.00
Alkalinity (μequiv L^{-1})	29	113.07	185.72	10.00	760.00
Taxon Richness*	222	2.07	0.97	1.00	5.00
PLEX	222	2.36	0.79	1.54	5.34

* *Number of submerged or floating plant taxa recorded in lake*

Summary description

Vegetation: Very species-poor. Dominated by *Sphagnum*; *Juncus bulbosus* frequently present.

Affinities: Equivalent to Type 1 in Palmer (1992) and Palmer *et al.* (1992).

Distribution: Small water bodies on peat or heathland, usually 100-350 m above sea level. Almost confined to Scotland, mainly in the western half of the country but a heavy concentration on the blanket bog of the Flow Country (Photo 1). The southerly outliers are: Abbot's Moss (Cheshire, basin mire), *Sphagnum* pools, Swanholme Pits (Lincolnshire heathland), Woolmer Ponds (Hampshire heathland), Little Sea Mere (Dorset heathland).

Chemistry: Highly acidic, low conductivity, very low alkalinity.

NVC: M1 (*Sphagnum auriculatum* bog pool).
M2 (*S. cuspidatum/recurvum* bog pool).

EC Habitats Directive: Natural dystrophic lakes are represented by pools on blanket bog in Scotland (e.g. in the Flow Country of Caithness and Sutherland) and on raised bogs (e.g. Abbots Moss, Cheshire).
Oligotrophic waters containing very few

minerals of sandy plains (*Littorelletalia uniflorae*) are represented by Little Sea, a shallow lake at Studland Dunes in south-west England.

Rare Species: The following species each occur in a single site in this group:
Apium inundatum Special Responsibility;
Eleogiton fluitans Special Responsibility;
Eriocaulon aquaticum Nationally Rare, Special Responsibility;
Nuphar pumila Nationally Scarce.

20 most constant macrophtes

Taxon	% of sites
Sphagnum (aquatic indet.)	**100.00**
Juncus bulbosus	**49.55**
Sparganium angustifolium	16.67
Potamogeton polygonifolius	7.66
Utricularia minor	7.21
Nymphaea alba	6.76
Glyceria fluitans	4.05
Potamogeton natans	1.80
Isoetes lacustris	1.35
Lemna minor	1.35
Sparganium natans	1.35
Callitriche stagnalis	0.90
Eriocaulon aquaticum	0.90
Fontinalis antipyretica	0.90
Lobelia dortmanna	0.90
Myriophyllum alterniflorum	0.90
Apium inundatum	0.45
Callitriche hamulata	0.45
Eleogiton fluitans	0.45
Elodea canadensis	0.45

Group B: Widespread, usually low-lying acid moorland or heathland pools and small lakes, with a limited range of plants, especially *Juncus bulbosus, Potamogeton polygonifolius* and *Sphagnum* spp.

No. of sites = 426

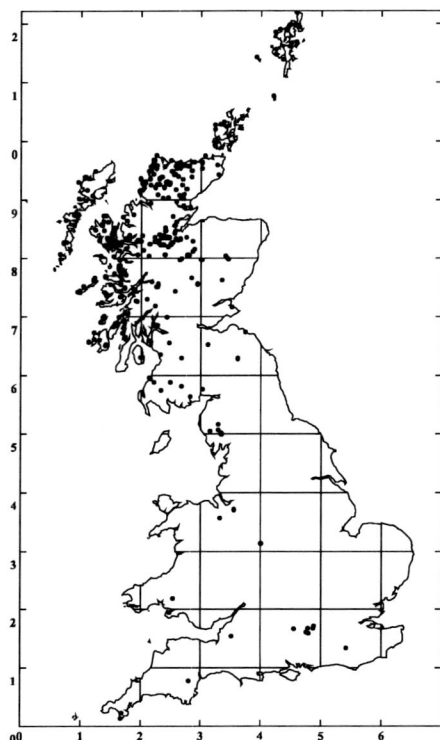

Summary features

Variable	n	Mean	Standard Deviation	Minimum	Maximum
Altitude	423	159.96	135.25	0.00	730.00
Surface area	422	1.95	7.09	0.01	96.19
pH	333	5.88	0.83	3.58	7.89
Conductivity (μS cm^{-1})	334	190.87	302.63	17.00	3000.00
Alkalinity (μequiv L^{-1})	110	484.08	1258.38	-8.00	10680.00
Taxon Richness*	426	5.26	2.98	1.00	22.00
PLEX	426	3.67	0.84	2.31	6.13

* Number of submerged or floating plant taxa recorded in lake

Summary description

Vegetation: Moderately species-poor. Typified by *P. polygonifolius, J. bulbosus* and *Sphagnum* spp.; *Nymphaea alba, Potamogeton natans* frequently present.

Affinities: Closest to Type 2 in Palmer (1992) and Palmer *et al.* (1992).

Distribution: Small waters on peat or heaths, usually <200 m above sea level. Mainly north and west Scotland and the Lake District. Southerly outliers include Broad Pool and Llyn Lech Owain (South Wales), Oak Mere (Cheshire; Photo 2), ponds on the Surrey and Hampshire heaths, Priddy Pool (Mendips), Stover Lake (Devon), pools on the Lizard heathland.

Chemistry: Acidic, low conductivity, low alkalinity.

NVC: A7 (*Nymphaea alba*), A9c (*Potamogeton natans: Juncus bulbosus-Myriophyllum alterniflorum* sub-community) and A24 (*Juncus bulbosus*).

EC Habitats Directive: Oligotrophic waters containing very few minerals of sandy plains (*Littorelletalia uniflorae*) are represented by Oak Mere (West Midlands of England). Hard oligo-mesotrophic waters with benthic vegetation of *Chara*

formations are represented by Croft Pascoe as a local variant of the habitat type within the Lizard Pools SAC.

Rare Species: *Apium inundatum* Special Responsibility (15 sites); *Eleogiton fluitans* Special Responsibility: this species is strongly associated with Group B (98 sites); *Eriocaulon aquaticum* Nationally Rare, Special Responsibility: Group B is the lake type most strongly associated with this species (22 sites); *Nuphar pumila* Nationally Scarce (17 sites); *Nymphoides peltata* Nationally Scarce (1 site); *Pilularia globulifera* Nationally Scarce, Species Action Plan (1 site); *Potamogeton filiformis* Nationally Scarce (1 site); *Ranunculus hederaceus* Special Responsibility (2 sites); *Ranunculus omiophyllus* Special Responsibility (1 site);

Note: The Pant-yr-Ochain Pools, Wales, are misclassified in this group because of the small number of plant records.

20 most constant macrophtes

Taxon	% of sites
Potamogeton polygonifolius	72.54
Juncus bulbosus	69.01
Sphagnum (aquatic indet.)	60.33
Nymphaea alba	45.07
Potamogeton natans	42.96
Eleogiton fluitans	23.00
Utricularia minor	22.30
Glyceria fluitans	17.14
Lobelia dortmanna	16.90
Sparganium angustifolium	16.67
Myriophyllum alterniflorum	14.08
Littorella uniflora	11.03
Sparganium natans	10.80
Chara spp.	10.56
Callitriche stagnalis	8.69
Utricularia intermedia sens. lat.	8.69
Nitella spp.	7.75
Sparganium emersum	5.87
Lemna minor	5.63
Utricularia vulgaris sens.lat.	5.40

Group C1: Northern, usually small to medium-sized, acid, largely mountain lakes, with a limited range of plants, but *Juncus bulbosus* and *Sparganium angustifolium* constant.

No. of sites = 256

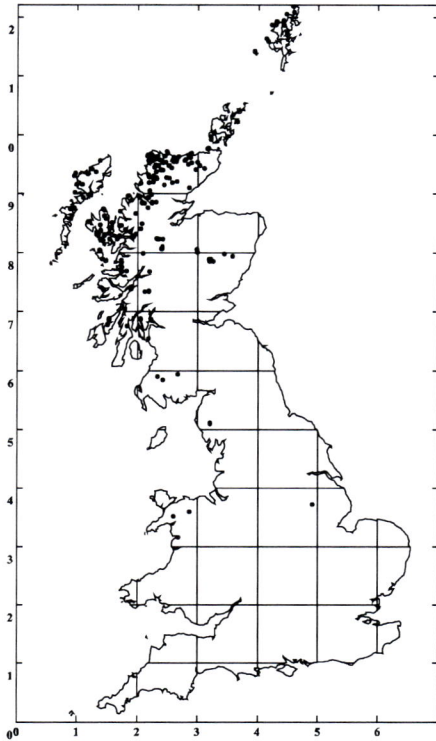

Summary features

Variable	n	Mean	Standard Deviation	Minimum	Maximum
Altitude	256	251.84	198.58	0.00	850.00
Surface area	255	4.85	16.24	0.10	193.00
pH	203	5.39	0.86	3.74	8.86
Conductivity (μS cm^{-1})	203	200.79	431.73	12.00	4000.00
Alkalinity (μequiv L^{-1})	47	69.42	155.83	-18.75	1020.00
Taxon Richness*	256	4.44	2.03	1.00	12.00
PLEX	256	3.70	0.52	2.69	5.25

* Number of submerged or floating plant taxa recorded in lake

Summary description

Vegetation: Species-poor. Typified by *Juncus bulbosus* and *Sparganium angustifolium*; *Sphagnum* frequently present sometimes in association with *Littorella uniflora*, *Lobelia dormanna* and *Isoetes lacustris*.

Affinities: Similar to Type 3 in Palmer (1992) and Palmer *et al.* (1992), but with *Sphagnum* more prominent, so tending towards Type 2.

Distribution: Generally small to medium-size upland waters, mostly on peat and 100-400 m above sea level. Almost exclusively in north and west Scotland. Southern outlier sites are a pit at Swanholme, Lincolnshire and Llyn Cau, at 470 m on Cadair Idris and Llyn Llagi, Snowdonia (Wales; Photo 3).

Chemistry: Acidic, low conductivity, very low alkalinity; some sites have very clear water.

NVC: A combination of A22 (*Littorella uniflora-Lobelia dortmanna*), A23 (*Isoetes lacustris/setacea*) and A24 (*Juncus bulbosus*).

EC Habitats Directive: Oligotrophic to mesotrophic standing waters with vegetation of the *Littorelletea uniflorae*

and/or of the *Isoëto-Nanojuncetea* are represented by Llyn Cau (North Wales).

Rare Species: *Eleogiton fluitans* Special Responsibility (7 sites); *Eriocaulon aquaticum* Nationally Rare, Special Responsibility (2 sites); *Nuphar pumila* Nationally Scarce (1 site).

Note: Swanholme Pits are located on heathland at low altitude.

20 most constant macrophtes

Taxon	% of sites
Juncus bulbosus	87.50
Sparganium angustifolium	65.63
Sphagnum (aquatic indet.)	54.69
Littorella uniflora	46.48
Lobelia dortmanna	37.89
Isoetes lacustris	28.13
Myriophyllum alterniflorum	21.48
Potamogeton polygonifolius	20.70
Fontinalis antipyretica	15.23
Glyceria fluitans	11.72
Potamogeton natans	10.94
Callitriche stagnalis	8.98
Callitriche hamulata	8.59
Subularia aquatica	6.25
Eleogiton fluitans	2.73
Utricularia minor	2.73
Potamogeton alpinus	2.34
Hippuris vulgaris	1.95
Sparganium natans	1.95
Isoetes echinospora	1.56

Group C2: North western, predominantly large, slightly acid, upland lakes, supporting a diversity of plant species, *Juncus bulbosus* constant, often with *Littorella uniflora* and *Lobelia dortmanna*, in association with *Myriophyllum alterniflorum*.

No. of sites = 1319

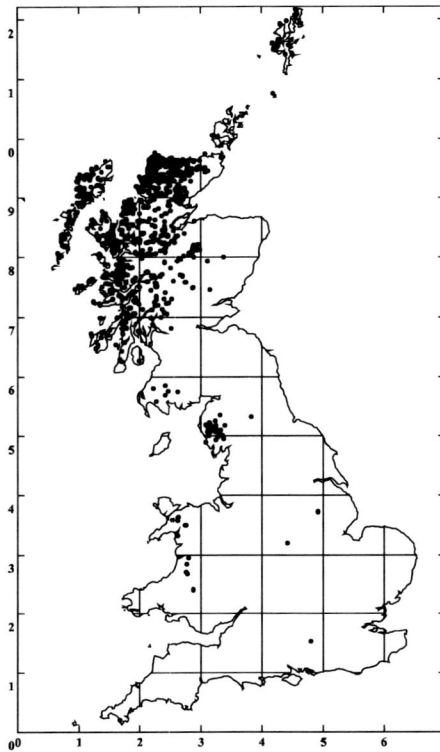

Summary features

Variable	n	Mean	Standard Deviation	Minimum	Maximum
Altitude	1315	167.36	129.08	0.00	730.00
Surface area	1314	20.49	97.04	0.05	1960.00
pH	1192	6.37	0.70	4.24	9.00
Conductivity (μS cm^{-1})	1194	129.54	172.88	12.50	2900.00
Alkalinity (μequiv L^{-1})	328	194.06	371.24	-8.75	2520.00
Taxon Richness*	1319	9.14	3.45	1.00	26.00
FLEX	1319	4.05	0.50	3.00	5.81

* Number of submerged or floating plant taxa recorded in lake

Summary description

Vegetation: Greater species diversity than C1. Typified by *Juncus bulbosus, Littorella uniflora, Lobelia dortmanna, Myriophyllum alterniflorum, Potamogeton polygonifolius* and *P. natans.*

Affinities: Equivalent to a mixture of Types 2 and 3 in Palmer (1992) and Palmer *et al.* (1992).

Distribution: A wide size range, including some large lakes, mostly below 250 m above sea level. A very heavy concentration in north and west Scotland, but very few in Orkney, eastern, central or southern Scotland (e.g. Loch an Eilein near Aviemore; Photo 4). Most of the Lake District tarns and lakes, and many upland lakes in Wales, are in this group. The only sites in southern England are a pit at Swanholme (Lincolnshire), Blackbrook Reservoir, (Leicestershire), ponds on Hampshire heathland and a pool on the Lizard.

Chemistry: Acidic, low conductivity, low alkalinity.

NVC: A combination of A22 (*Littorella uniflora-Lobelia dortmanna*), A23 (*Isoetes lacustris/setacea*), A24 (*Juncus bulbosus*), A7 (*Nymphaea alba*), A9c (*Potamogeton natans*) and A14 (*Myriophyllum alterniflorum*). A13

(*Potamogeton perfoliatus-Myriophyllum alterniflorum*) may be present in the more enriched situations.

EC Habitats Directive: Oligotrophic waters containing very few minerals of sandy plains (*Littorelletalia uniflorae*) are represented by several sites located on the South Uist machair (Lochs Fada, Cuithe Moire, a'Phuirt-ruaidh, a'Chnoic Bhuidha, Schoolhouse Loch). Oligotrophic to mesotrophic standing waters with vegetation of the *Littorelletea uniflorae* and/or of the *Isoëto-Nanojuncetea* are represented by Llyn Cwellyn and Llyn Idwal (Snowdonia), Loch Einich (Cairngorms), Loch Achtriochtan (Glen Coe), Wast Water and tarns in the high fells of the Lake District. Hard oligo-mesotrophic waters with benthic vegetation of *Chara* spp. are represented by Loch Cill Chriosd (Isle of Skye). It is possible that natural dystrophic lakes with diverse macrophyte assemblages may occur in this group.

Rare Species: *Apium inundatum* Special Responsibility (32 sites); *Eleogiton fluitans* Special Responsibility: Group C2 is the lake type most strongly associated with this species (371 sites); *Eriocaulon aquaticum* Nationally Rare,

20 most constant macrophtes

Taxon	% of sites
Juncus bulbosus	92.57
Littorella uniflora	88.40
Lobelia dortmanna	83.55
Potamogeton polygonifolius	77.79
Myriophyllum alterniflorum	77.03
Potamogeton natans	70.66
Sparganium angustifolium	48.45
Isoetes lacustris	44.96
Nymphaea alba	35.78
Eleogiton fluitans	28.13
Glyceria fluitans	26.99
Fontinalis antipyretica	26.31
Sphagnum (aquatic indet.)	22.74
Nitella spp.	21.53
Subularia aquatica	20.17
Chara spp.	18.57
Utricularia minor	16.60
Utricularia intermedia sens. lat.	15.09
Callitriche hamulata	11.90
Potamogeton perfoliatus	10.54

Summary description (cont)

Special Responsibility (22 sites);
Luronium natans EC Habitats
Directive/Bern Convention, Schedule 8,
Nationally Scarce, Special Responsibility,
Species Action Plan (3 sites);
Najas flexilis EC Habitats Directive/Bern
Convention, Schedule 8, Nationally
Scarce, Species Action Plan (7 sites);
Nuphar pumila Nationally Scarce
(17 sites);
Nymphoides peltata Nationally Scarce
(2 sites);
Pilularia globulifera Nationally Scarce,
Species Action Plan (9 sites);
Potamogeton coloratus Nationally Scarce
(1 site);
Potamogeton epihydrus Vulnerable, Species
Action Plan: Group C2 contains the only
site (in Skye) in the database for this
species; record to be confirmed;
Potamogeton filiformis Nationally Scarce
(7 sites);
Potamogeton rutilus Nationally Rare,
Species Action Plan (1 site);
Ranunculus hederaceus Special
Responsibility (1 site);
Ranunculus omiophyllus Special
Responsibility (5 sites);
Chara curta Nationally Scarce, Special
Responsibility, Species Action Plan
(2 sites);
Nitella confervacea Nationally Rare: this
species is strongly associated with Group
C2 (5 sites);
Nitella gracilis Vulnerable, Species Action
Plan (2 sites).

Note: More than 40% of the lakes in the
dataset are in this group and they cover a
wide range of lake habitat types, as
illustrated by the links made with the
NVC and Habitats Directive. Water colour
may vary from dystrophic brown water to
clear oligotrophic water. The classic
oligotrophic waters are in this group.
Examples are Loch Ard (Trossachs), East
Loch Ollay (South Uist), Buttermere, Wast
Water and Hodson's Tarn (Lake District,
England) and Llyn Idwal (Wales). Tarn
Dub in upper Teesdale appears to be
misclassified in this group because of the
lack of *Chara* records in the survey data.

Group D: Widespread, often large, mid-altitude circumneutral lakes, with a high diversity of plants, including *Littorella uniflora*, *Myriophyllum alterniflorum*, *Callitriche hamulata*, *Fontinalis antipyretica* and *Glyceria fluitans*.

No. of sites = 370

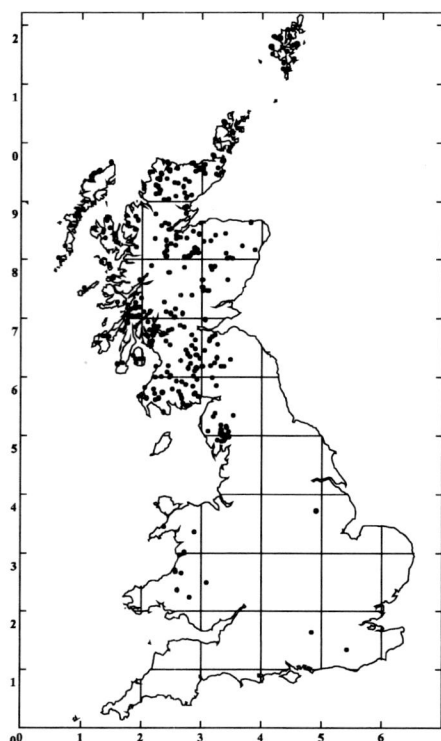

Summary features

Variable	n	Mean	Standard Deviation	Minimum	Maximum
Altitude	370	174.38	163.30	0.00	840.00
Surface area	368	84.46	437.54	0.01	5640.00
pH	340	6.70	0.75	4.20	9.23
Conductivity (μS cm^{-1})	337	189.47	355.50	16.80	5400.00
Alkalinity (μequiv L^{-1})	177	501.49	530.03	6.00	4000.00
Taxon Richness*	370	10.31	5.47	1.00	30.00
FLEX	370	5.37	0.63	3.85	7.02

* *Number of submerged or floating plant taxa recorded in lake*

Summary description

Vegetation: A species-rich group. Typified by *Littorella uniflora*, *Myriophyllum alterniflorum*, *Callitriche hamulata*, *Fontinalis antipyretica* and *Glyceria fluitans*. The most constant pondweeds are *P. natans*, *P. polygonifolius*, *P. perfoliatus* and *P. berchtoldii*. *Nitella* spp. and, to a lesser extent, *Chara* spp. are represented.

Affinities: Contains many of the sites in Type 5A in Palmer (1992) and Palmer *et al.* (1992) together with some Type 3 and Type 4 waters.

Distribution: This lake group has a wide size range, including some large lakes, mostly < 250 m above sea level. Numerous and scattered throughout Scotland, frequent in the Lake District e.g. Bassenthwaite Lake; Photo 5) and Wales. Sites are very rare in southern England and include two pits at Swanholme (Lincolnshire), and ponds on Berkshire heathland.

Chemistry: Weakly acidic, low conductivity and moderate alkalinity.

NVC: A combination of A22 (*Littorella uniflora-Lobelia dortmanna*), A23 (*Isoetes lacustris/setacea*), A13 (*Potamogeton perfoliatus-Myriophyllum alterniflorum*), A7 (*Nymphaea alba*), A9c

(*Potamogeton natans: Juncus bulbosus-Myriophyllum alterniflorum* sub-community) and A15 (*Elodea canadensis*); in more nutrient-poor areas probably also A14 (*Myriophyllum alterniflorum*) and A24 (*Juncus bulbosus*).

EC Habitats Directive: Oligotrophic to mesotrophic standing waters with vegetation of the *Littorelletea uniflorae* and/or of the *Isoëto-Nanojuncetea* are represented by Loch of Clunie and Loch Maree (Scottish Highlands), Loch Insh (River Spey), Loch Kinord (Dee Valley, Scotland), Bassenthwaite Lake (English Lake District).

Rare Species: *Apium inundatum* Special Responsibility (48 sites); *Elatine hydropiper* Nationally Scarce (4 sites); *Eleogiton fluitans* Special Responsibility (26 sites); *Luronium natans* EC Habitats Directive/ Bern Convention, Schedule 8, Nationally Scarce, Special Responsibility, Species Action Plan: Group D is the type most strongly associated with this species (5 sites); *Najas flexilis* EC Habitats Directive/Bern Convention, Schedule 8, Nationally Scarce, Species Action Plan (4 sites); *Nuphar pumila* Nationally Scarce: Group D is the type most strongly associated with this species (17 sites);

20 most constant macrophtes

Taxon	% of sites
Littorella uniflora	70.00
Callitriche hamulata	66.76
Glyceria fluitans	64.05
Myriophyllum alterniflorum	62.70
Fontinalis antipyretica	61.35
Nitella spp.	57.57
Potamogeton natans	53.78
Juncus bulbosus	52.70
Sparganium angustifolium	43.24
Potamogeton polygonifolius	35.41
Callitriche stagnalis	34.59
Potamogeton berchtoldii	33.51
Isoetes lacustris	32.16
Nymphaea alba	24.05
Chara spp.	21.89
Potamogeton perfoliatus	21.35
Elodea canadensis	20.81
Lobelia dortmanna	20.81
Potamogeton obtusifolius	16.49
Potamogeton alpinus	16.22

Summary description (cont)

Pilularia globulifera Nationally Scarce, Species Action Plan: Group D is the type most strongly associated with this species (10 sites);

Potamogeton coloratus Nationally Scarce (1 site);

Potamogeton compressus Nationally Scarce, Species Action Plan: Group D contains the only site (Loch of Aboyne) in the database for this species;

Potamogeton filiformis Nationally Scarce (2 sites);

Ranunculus hederaceus Special Responsibility (17 sites);

Ranunculus omiophyllus Special Responsibility (3 sites);

Nitella gracilis Vulnerable, Species Action Plan (1 site).

Note: This group includes classic mesotrophic sites such as Loch Insh, Lochs Clunie, Marlee and Craiglush (Perth, Scotland), the Lake of Menteith (Stirling), Bassenthwaite Lake and Windermere (The Lake District), Llyn Eiddwen and Llyn Fanod (mid-Wales) and Bala Lake/Llyn Tegid (north Wales). Llyn Y Fan Fawr (Brecon Beacons, Wales) is probably misclassified in this group as a result of the small number of plant records for this site.

Group E: Northern, often large, low altitude and coastal, above-neutral lakes with high diversity of plant species, including *Littorella uniflora, Myriophyllum alterniflorum, Potamogeton perfoliatus* and *Chara* spp.

No. of sites = 186

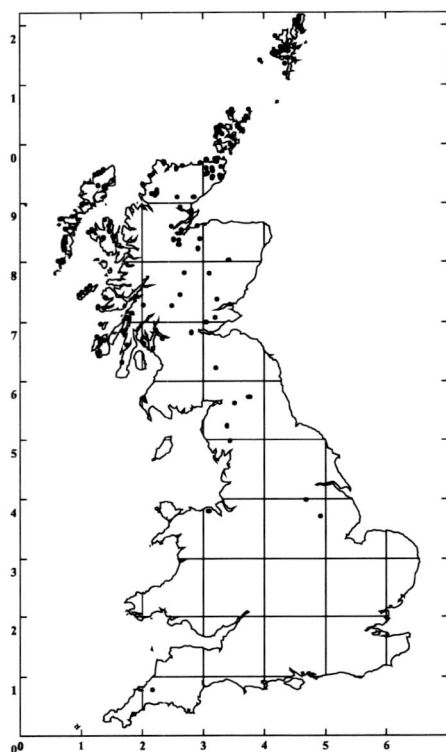

Summary features

Variable	n	Mean	Standard Deviation	Minimum	Maximum
Altitude	184	70.31	122.30	0.00	750.00
Surface area	179	30.01	76.73	0.01	867.80
pH	153	7.75	0.79	6.00	9.80
Conductivity (μS cm^{-1})	153	434.51	455.27	39.00	3720.00
Alkalinity (μequiv L^{-1})	71	926.48	725.72	100.00	4360.00
Taxon Richness*	186	11.85	5.47	2.00	32.00
PLEX	186	6.00	0.55	4.83	7.60

* *Number of submerged or floating plant taxa recorded in lake*

Summary description

Vegetation: A species-rich group. Typified by *Littorella uniflora, Myriophyllum alterniflorum, Potamogeton perfoliatus* and *Chara*. Other common pondweeds are *P. natans, P. gramineus, P. berchtoldii, P. polygonifolius* and *P. filiformis*.

Affinities: Closest to Type 4 in Palmer (1992) and Palmer et al. (1992), but some similarities to Type 3 and 5A.

Distribution: A wide size range, including some large lakes, mostly < 100 m above sea level. Largely coastal lakes in north and west Scotland, especially on the islands (e.g. Loch of Hundland, mainland Orkney; Photo 6); scattered sites inland. Rare in the Lake District (e.g. Ullswater) and Wales (only Llyn Helyg). Sites are very rare in southern England and include four pits at Swanholme (Lincolnshire).

Chemistry: Circumneutral and above pH, moderate conductivity and alkalinity.

NVC: A combination of A22 (*Littorella uniflora-Lobelia dortmanna*), A23 (*Isoetes lacustris/setacea*), A13 (*Potamogeton perfoliatus-Myriophyllum alterniflorum*), A9c (*Potamogeton natans: Juncus bulbosus-Myriophyllum alterniflorum* sub-community) and A10 (*Polygonum amphibium*). In more nutrient-poor areas

of these waters A14 (*Myriophyllum alterniflorum*) and A24 (*Juncus bulbosus*) probably occur.

EC Habitats Directive: Oligotrophic to mesotrophic standing waters with vegetation of the *Littorelletea uniflorae* and/or of the *Isoëto-Nanojuncetea* are represented by Loch nan Cat (Ben Lawers), Loch Ussie (Scottish Highlands), Loch Davan (Dee Valley), Ullswater (Lake District).
Hard oligo-mesotrophic waters with benthic vegetation of *Chara* spp. are represented by the Scottish Durness cluster of three marl lochs (Croispol, Borralie and Caladail), several lochs in the South Uist machair (Loch Hallan, Grogarry Loch, Loch an Eilean, Mid Loch Ollay and Loch Toronish).
Natural eutrophic lakes with *Magnopotamion* or *Hydrocharition*-type vegetation are represented by Loch of Wester, Loch Roag, Loch na Liana Moire, Loch of Isbister (Scotland), and Greenlee and Broomlee Loughs (Northumberland). Coastal lagoons are represented by a complex at Loch Roag on the Western Isles.

Rare Species: *Apium inundatum* Special Responsibility: Group E is the type most strongly associated with this species (45 sites);

20 most constant macrophtes

Taxon	% of sites
Littorella uniflora	88.71
Myriophyllum alterniflorum	83.87
Chara spp.	77.42
Potamogeton perfoliatus	68.82
Potamogeton natans	58.60
Potamogeton gramineus	54.84
Fontinalis antipyretica	50.00
Glyceria fluitans	46.77
Juncus bulbosus	44.09
Potamogeton filiformis	41.40
Nitella spp.	38.17
Potamogeton polygonifolius	37.63
Sparganium angustifolium	32.26
Potamogeton berchtoldii	30.65
Callitriche stagnalis	29.03
Potamogeton gramineus x perfoliatus	26.88
Apium inundatum	24.19
Callitriche hermaphroditica	23.66
Isoetes lacustris	21.51
Persicaria amphibia	20.43

Summary description (cont)

Eleogiton fluitans Special Responsibility (17 sites);

Najas flexilis EC Habitats Directive/Bern Convention, Schedule 8, Nationally Scarce, Species Action Plan: Group E is the type most strongly associated with this species (11 sites);

Nymphoides peltata Nationally Scarce (1 site);

Pilularia globulifera Nationally Scarce, Species Action Plan (4 sites);

Potamogeton coloratus Nationally Scarce, (1 site);

Potamogeton filiformis Nationally Scarce: Group E is the type most closely associated with this species (77 sites);

Potamogeton friesii Nationally Scarce (3 sites);

Potamogeton rutilus Nationally Rare, Species Action Plan: Group E is the type most strongly associated with this species (7 sites);

Ranunculus hederaceus Special Responsibility (7 sites);

Ruppia cirrhosa Nationally Scarce (2 sites);

Chara aculeata (*pedunculata*) Nationally Scarce: the only site in the database for this species is Loch Ballygrant (Island of Islay);

Chara curta Nationally Scarce, Special Responsibility, Species Action Plan: this species is strongly associated with Group E (5 sites);

Nitella confervacea Nationally Rare (2 sites);

Nitella mucronata Nationally Scarce: the only site in the database for this species is in Shetland.

Note: Classic coastal, moderately to strongly calcareous marl sites are included here, such as Loch Eye, and some machair lochs in South Uist, Coll and Tiree (e.g. Mid Loch Ollay, Loch Hallan, Loch Roag). At Swanholme, some of the pits that appear to be fed by calcareous groundwater are in this group. Broomlee and Greenlee Loughs, (Northumberland) are other inland examples.

Group F: Widespread, usually medium-sized, lowland, above neutral lakes, with a limited range of species, but typified by water-lilies and other floating-leaved vegetation.

No. of sites = 48

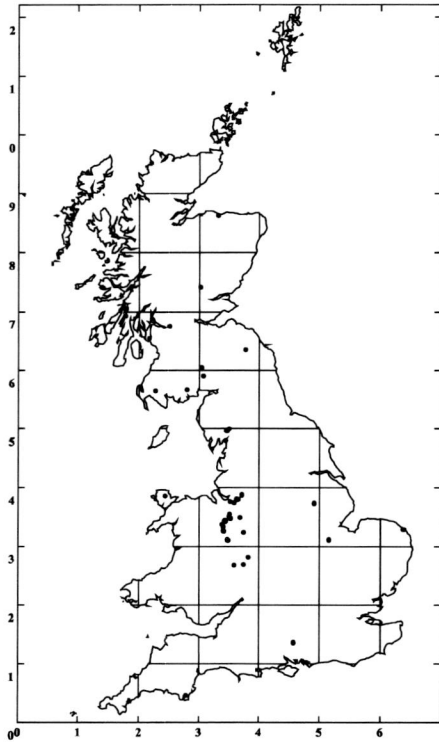

Summary features

Variable	n	Mean	Standard Deviation	Minimum	Maximum
Altitude	48	66.15	39.41	0.00	180.00
Surface area	47	10.03	16.03	0.25	75.34
pH	41	7.61	0.77	4.51	8.80
Conductivity (μS cm^{-1})	43	537.14	800.37	87.00	5500.00
Alkalinity (μequiv L^{-1})	11	1529.77	991.66	520.00	3997.25
Taxon Richness*	48	5.23	2.29	1.00	13.00
PLEX	48	6.97	0.98	4.81	8.28

* Number of submerged or floating plant taxa recorded in lake

Summary description

Vegetation: A species-poor group. Typified by *Nuphar lutea*, accompanied by *Nymphaea alba*, *Lemna minor*, *Callitriche stagnalis* and *Persicaria amphibia*.

Affinities: Nearest to Type 9 in Palmer (1992) and Palmer *et al.* (1992), but some similarities to Type 8.

Distribution: Mostly medium-sized lakes < 100 m above sea level. Centred on the West Midland Meres, but with a few outliers in Scotland, Wales and the rest of England (e.g. Slapton Ley, Devon; Photo 7).

Chemistry: Above neutral pH, moderate conductivity, high alkalinity.

NVC: A8b and c (*Nuphar lutea: Callitriche stagnalis-Zanichellia palustris* and *Nymphaea* sub-communities).

EC Habitats Directive: Hard oligo-mesotrophic waters with benthic vegetation of *Chara* formations are represented by Llyn yr Wyth Eidion (Anglesey) and Hawes Water (Silverdale, northern England).

Rare Species: *Nuphar pumila* Nationally Scarce (1 site); *Pilularia globulifera* Nationally Scarce,

Species Action Plan (1 site); *Ranunculus hederaceus* Special Responsibility (2 sites); *Ranunculus omiophyllus* Special Responsibility (2 sites).

Note: This group contains classic water-lily sites, including Llyn yr Wyth Eidion (Anglesey, Wales), and the West Midlands Meres that have water-lily vegetation.

20 most constant macrophtes

Taxon	% of sites
Nuphar lutea	89.58
Lemna minor	60.42
Nymphaea alba	50.00
Callitriche stagnalis	47.92
Persicaria amphibia	41.67
Zannichellia palustris	31.25
Elodea canadensis	20.83
Glyceria fluitans	12.50
Lemna trisulca	12.50
Potamogeton pectinatus	12.50
Hippuris vulgaris	10.42
Potamogeton berchtoldii	10.42
Sphagnum spp.	10.42
Callitriche platycarpa	8.33
Potamogeton natans	8.33
Callitriche hamulata	6.25
Chara spp.	6.25
Potamogeton pusillus	6.25
Sparganium natans	6.25
Callitriche obtusangula	4.17

Group G: Central and eastern, above neutral, lowland lakes, with *Lemna minor, Elodea canadensis, Potamogeton natans* and *Persicaria amphibia.*

No. of sites = 281

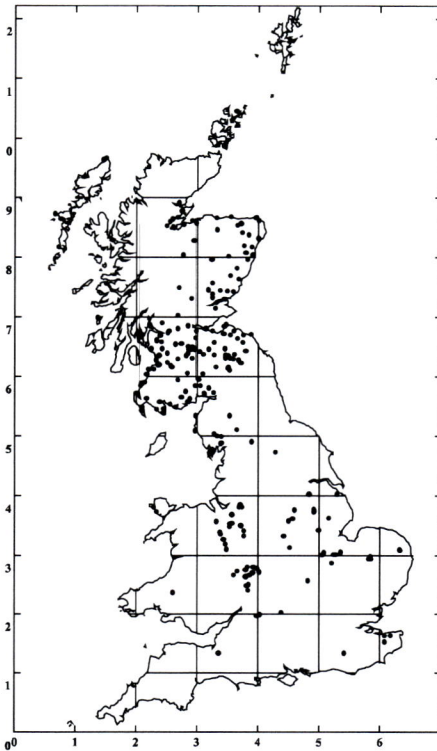

20 most constant macrophtes

Taxon	% of sites
Lemna minor	75.44
Persicaria amphibia	46.98
Elodea canadensis	44.84
Potamogeton natans	42.70
Glyceria fluitans	40.93
Callitriche stagnalis	39.50
Potamogeton berchtoldii	30.60
Callitriche hamulata	27.05
Potamogeton crispus	26.33
Potamogeton obtusifolius	25.98
Lemna trisulca	24.56
Chara spp.	22.06
Myriophyllum spicatum	20.64
Potamogeton pusillus	16.73
Nitella spp.	16.37
Nuphar lutea	16.01
Nymphaea alba	16.01
Potamogeton pectinatus	15.66
Zannichellia palustris	14.23
Fontinalis antipyretica	12.10

Summary features

Variable	n	Mean	Standard Deviation	Minimum	Maximum
Altitude	270	76.22	82.06	0.00	370.00
Surface area	241	6.54	16.46	0.05	182.00
pH	216	7.45	0.78	5.10	9.86
Conductivity (μS cm^{-1})	194	572.01	1287.80	61.00	15420.00
Alkalinity (μequiv L^{-1})	154	1833.67	1515.64	40.00	10200.00
Taxon Richness*	281	7.71	4.19	1.00	27.00
PLEX	281	7.29	0.68	5.62	8.85

* Number of submerged or floating plant taxa recorded in lake

Summary description

Vegetation: A moderately species-rich group. Typified by *Lemna minor, Elodea canadensis, Potamogeton natans* and *Persicaria amphibia.*

Affinities: Nearest to Type 10A in Palmer (1992) and Palmer *et al.* (1992), but with elements of Type 8.

Distribution: Mostly small to medium-sized lakes < 100 m above sea level. A widespread group, well represented in lowland England, especially the West Midlands Meres (e.g. The Mere, Ellesmere; Photo 8), common in south and east Scotland but rare in Wales.

Chemistry: Circumneutral pH, moderate conductivity, high alkalinity.

NVC: A2 (*Lemna minor*), A9b (*Potamogeton natans: Elodea canadensis* sub-community), A10 (*Polygonum amphibium*) and A15 (*Elodea canadensis*).

EC Habitats Directive: Natural eutrophic lakes with *Magnopotamion* or *Hydrocharition*-type vegetation are represented by Llyn Coron (Anglesey, Wales).

Rare Species: *Alisma gramineum* Schedule 8, Critically Endangered: the only site for this species (Westwood Great Pool, Worcestershire) is in Group G; *Apium inundatum* Special responsibility (18 sites); *Callitriche truncata* Nationally Scarce: the only site for this species (Clumber Park Lake) is in Group G; *Elatine hydropiper* Nationally Scarce: Group G is the type most strongly associated with this species (9 sites); *Nuphar pumila* Nationally Scarce (2 sites); *Nymphoides peltata* Nationally Scarce (3 sites); *Potamogeton coloratus* Nationally Scarce (2 sites); *Potamogeton filiformis* Nationally Scarce (1 site); *Potamogeton friesii* Nationally Scarce (4 sites); *Ranunculus hederaceus* Special Responsibility (25 sites); *Ranunculus omiophyllus* Special Responsibility (2 sites); *Ruppia cirrhosa* Nationally Scarce (1 site); *Wolffia arrhiza* Nationally Scarce: the only site for this species (Southlake Moor, Somerset) is in Group G; *Chara rudis* Nationally Rare (1 site).

Note: This is the commonest lowland eutrophic lake type. Well-known sites in England include Semer Water, some of the Cotswold Water Park Pits and Sunbiggin

33

Summary description (cont)

Tarn (an upland site). In Wales, Llyn Coron and Llyn Penrhyn cluster together on Anglesey, while the Lower Talley Lake is a southern outlier in Wales. Scottish sites include Castle Loch (Dumfries) and Hen Poo (Borders). Some sites are known to be enriched and the natural vegetation community may be distorted by the presence of *Elodea canadensis*. A number of marl lakes (e.g. Cotswold Water Park Pits) are included within this group.

Group H: Northern, small, circumneutral, lowland lakes, with low species diversity characterised by the presence of *Glyceria fluitans* and *Callitriche stagnalis*.

No. of sites = 101

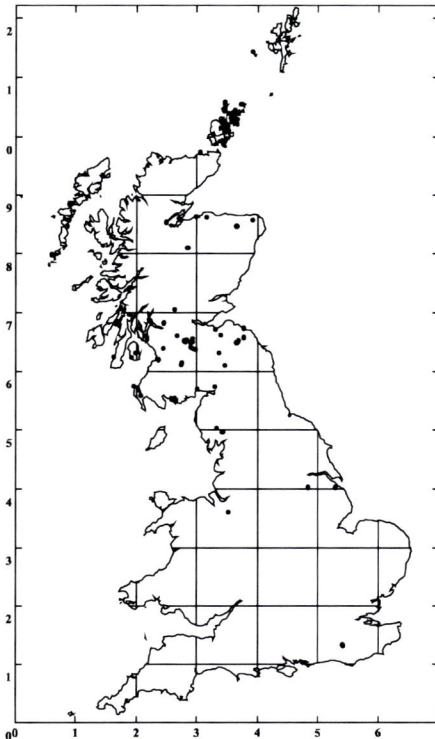

Summary features

Variable	n	Mean	Standard Deviation	Minimum	Maximum
Altitude	101	70.10	84.81	0.00	330.00
Surface area	100	3.64	21.04	0.01	205.80
pH	55	7.05	1.08	3.90	9.00
Conductivity (μS cm^{-1})	56	880.16	2780.26	83.00	21000.00
Alkalinity (μequiv L^{-1})	40	1583.25	1492.61	100.00	6800.00
Taxon Richness*	101	2.83	1.51	1.00	7.00
PLEX	101	7.03	0.65	4.62	8.04

* Number of submerged or floating plant taxa recorded in lake

Summary description

Vegetation: A very species-poor group. Typified by *Glyceria fluitans* and *Callitriche stagnalis*.

Affinities: No near counterpart in Palmer (1992) and Palmer *et al.* (1992), but elements of Type 8.

Distribution: Mostly small water bodies < 100 m above sea level. An almost exclusively lowland Scottish lake type, with a heavy concentration in eastern Orkney (e.g. Mill Loch, Eday, Orkney; Photo 9). Southerly outliers are Blelham Fish Pond (Lake District), Peckforton Mere (West Midlands) and Pippingford Park Lakes (south-east England). No representative in Wales.

Chemistry: Circumneutral pH, moderate conductivity, high alkalinity.

NVC: A16 (*Callitriche stagnalis*).

EC Habitats Directive: None.

Rare Species: *Apium inundatum* Special responsibility (11 sites); *Potamogeton filiformis* Nationally Scarce (2 sites); *Ranunculus hederaceus* Special Responsibility (6 sites); *Ranunculus omiophyllus* Special Responsibility (1 site).

Note: A closer investigation of the sites on Orkney may provide an insight into the environmental characteristics of this lake group.

20 most constant macrophtes

Taxon	% of sites
Callitriche stagnalis	**74.26**
Glyceria fluitans	**70.30**
Lemna minor	14.85
Callitriche hamulata	13.86
Apium inundatum	10.89
Ranunculus baudotii	10.89
Ranunculus aquatilis sens. str.	9.90
Fontinalis antipyretica	7.92
Ranunculus peltatus	7.92
Potamogeton natans	6.93
Hippuris vulgaris	5.94
Ranunculus hederaceus	5.94
Nitella spp.	4.95
Potamogeton polygonifolius	3.96
Callitriche platycarpa	2.97
Potamogeton berchtoldii	2.97
Sphagnum (aquatic indet.)	2.97
Callitriche hermaphroditica	1.98
Chara spp.	1.98
Potamogeton crispus	1.98

Group I: Widespread, mostly moderately large, base-rich lowland lakes, with *Chara* spp., *Myriophyllum spicatum* and a diversity of *Potamogeton* species.

No. of sites = 203

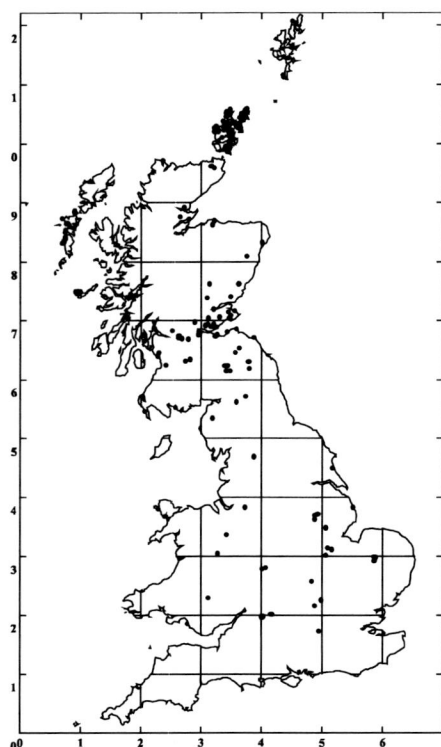

Summary features

Variable	n	Mean	Standard Deviation	Minimum	Maximum
Altitude	192	42.73	69.99	0.00	390.00
Surface area	184	23.79	109.91	0.01	1415.30
pH	163	8.39	0.91	6.00	10.80
Conductivity (μS cm^{-1})	160	1045.25	2134.66	40.00	15400.00
Alkalinity (μequiv L^{-1})	71	1790.29	837.48	400.00	5880.00
Taxon Richness*	203	9.12	4.80	1.00	22.00
PLEX	203	7.61	0.51	6.13	8.85

* Number of submerged or floating plant taxa recorded in lake

20 most constant macrophtes

Taxon	% of sites
Chara spp.	68.97
Myriophyllum spicatum	59.11
Potamogeton pectinatus	53.69
Potamogeton pusillus	46.80
Callitriche stagnalis	38.42
Potamogeton filiformis	37.93
Potamogeton perfoliatus	34.98
Zannichellia palustris	34.98
Potamogeton crispus	33.99
Callitriche hermaphroditica	32.51
Persicaria amphibia	32.51
Glyceria fluitans	31.03
Potamogeton natans	28.57
Lemna minor	28.08
Littorella uniflora	27.59
Potamogeton berchtoldii	27.09
Elodea canadensis	26.60
Ranunculus baudotii	24.63
Hippuris vulgaris	18.72
Fontinalis antipyretica	17.73

Summary description

Vegetation: A moderately species-rich group. Typified by *Chara*, *Myriophyllum spicatum*, *Callitriche stagnalis* and *C. hermaphroditica*, *Zannichellia palustris* and a wide range of pondweeds - *P. filiformis*, *P. pectinatus*, *P. pusillus*. *P. filiformis* is characteristic of the Scottish coastal sites.

Affinities: A mixture of Types 10A and 7 in Palmer (1992) and Palmer et al. (1992).

Distribution: Widespread distribution. Mostly medium-sized water bodies < 75 m above sea level. In Scotland, there are concentrations in Orkney, on the machair of the Outer Hebrides and Tiree, and the Central Lowlands. Uncommon but widespread in England and Wales.

Chemistry: Relatively high pH, moderate conductivity, high alkalinity.

NVC: A11 (*Potamogeton pectinatus-Myriophyllum spicatum*) and probably all three sub-communities (*P. pusillus, Elodea canadensis* and *P. filiformis*) occur. Some possible affinities with A16 (*Callitriche stagnalis*).

EC Habitats Directive: Hard oligo-mesotrophic waters with benthic vegetation of *Chara* formations are represented by Malham Tarn (northern Yorkshire, England; Photo 10), and Kenfig Pool (South Wales), Loch Baile a'Ghobhainn and Loch Fiart (Island of Lismore, Argyll, Scotland). Natural eutrophic lakes with *Magnopotamion* or *Hydrocharition*-type vegetation are represented by Llangorse Lake and Llyn Dinam (Wales), Crag Lough (Northumberland), Loch a'Phuill, Loch Achnacloich, Loch Watten and Loch Bhasapol (Scotland) and lochs in the Outer Hebrides (Loch nam Feithean, West Loch Ollay, Loch Ardvule, Loch Stilligarry).

Rare Species: *Apium inundatum* Special Responsibility (15 sites); *Elatine hydropiper* Nationally Scarce (2 sites); *Eleogiton fluitans* Special Responsibility (1 site); *Nuphar pumila* Nationally Scarce (2 sites); *Nymphoides peltata* Nationally Scarce (4 sites); *Potamogeton coloratus* Nationally Scarce, Special Responsibility: Group I is the lake type most strongly associated with this species (4 sites); *Potamogeton filiformis* Nationally Scarce: this species is strongly associated with Group I (77 sites); *Potamogeton friesii* Nationally Scarce:

Summary description (cont)

Group I is the type most strongly associated with this species (21 sites); *Ranunculus hederaceus* Special Responsibility (10 sites); *Ranunculus omiophyllus* Special Responsibility (4 sites); *Ruppia cirrhosa* Nationally Scarce (1 site); *Chara rudis* Nationally Rare (1 site).

Note: This group seems to be a mixture of coastal and inland calcareous lakes. For instance, in Wales, this group is represented by Kenfig Pool lying within an extensive coastal sand-dune system, while Llangorse Lake is located inland on a tributary of the River Wye. In England, Hornsea Mere occurs on the coast, and contrasts with the location of most of the Cotswold Water Park lakes, several Breckland Meres, Crag Lough (Roman Wall), Malham Tarn (Yorkshire) and two pits at Swanholme (Lincolnshire). In Scotland, the coastal lochs in South Uist and Loch Lanlish (Durness) differ from the more inland locations of Lochs Leven, Branxholme Easter (Borders), and Watten (Caithness). Many of the sites in this group are marl lakes on limestone, chalk or machair.

Group J: Northern coastal, brackish lakes, with *Potamogeton pectinatus, Enteromorpha* spp., *Ruppia maritima* and fucoid algae.

No. of sites = 35

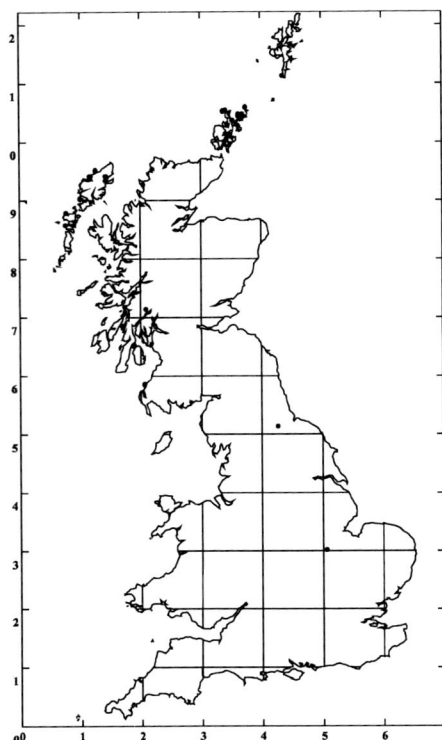

Summary features

Variable	n	Mean	Standard Deviation	Minimum	Maximum
Altitude	35	3.43	9.68	0.00	40.00
Surface area	34	3.18	5.19	0.10	26.00
pH	28	8.14	0.80	5.90	9.05
Conductivity (μS cm^{-1})	28	20812.46	27107.52	59.00	110000.00
Alkalinity (μequiv L^{-1})	15	1702.40	909.33	80.00	3400.00
Taxon Richness*	35	3.17	2.33	1.00	9.00
PLEX	35	7.71	0.92	4.23	8.85

* Number of submerged or floating plant taxa recorded in lake

Summary description

Vegetation: A very species-poor group. Typified by *Enteromorpha, Ruppia maritima, Potamogeton pectinatus, Callitriche stagnalis,* and sometimes fucoid algae.

Affinities: Equivalent to Type 6, but many of the brackish sites in Palmer (1992) and Palmer *et al.* (1992) were not included in the JNCC dataset.

Distribution: Small to medium-sized water bodies at sea level. Exclusively coastal, occurring in Shetland, Orkney (the main concentration; e.g. Loch of the Stack, Westray; Photo 11), the Outer Hebrides and the west coast of mainland Scotland.

Chemistry: High pH, conductivity and alkalinity.

NVC: A12 (*Potamogeton pectinatus*).

EC Habitats Directive: The vegetation of coastal lagoons may include *Zostera* spp., *Ruppia* spp., *Potamogeton* spp. and stoneworts.

Rare Species: *Apium inundatum* Special Responsibility (1 site); *Eleogiton fluitans* Special Responsibility (1 site); *Potamogeton filiformis* Nationally Scarce (2 sites);

Ranunculus hederaceus Special Responsibility (4 sites).

Note: Two sites - one of the Stibbington Gravel pits (only records *Nuphar lutea* and *Potamogeton pectinatus*) and Hell Kettles (only records *Lemna minor, Potamogeton pectinatus* and *Ranunculus circinatus*) - are misclassified as J because of their extremely poor flora.

20 most constant macrophtes

Taxon	% of sites
Ruppia maritima	60.00
Enteromorpha spp.	54.29
Potamogeton pectinatus	54.29
Callitriche stagnalis	22.86
Fucoid algae	22.86
Glyceria fluitans	14.29
Chara spp.	11.43
Ranunculus baudotii	11.43
Ranunculus hederaceus	11.43
Potamogeton natans	8.57
Potamogeton polygonifolius	8.57
Potamogeton filiformis	5.71
Zannichellia palustris	5.71
Apium inundatum	2.86
Callitriche hamulata	2.86
Eleogiton fluitans	2.86
Juncus bulbosus	2.86
Lemna minor	2.86
Littorella uniflora	2.86
Myriophyllum spicatum	2.86

Photo 1. Dystrophic lakes in the Flow Country, northern Scotland; example of Group A (photograph from Steve Moore, Scottish Natural Heritage).

Photo 2. Oak Mere, Cheshire, England; example of Group B (photograph from Chris Walker, English Nature).

Photo 3. Llyn Llagi, Snowdonia, Wales; example of Group C1 (photograph from Catherine Duigan, Countryside Council for Wales).

Photo 4. Loch an Eilein near Aviemore, Scotland; example of Group C2 (photograph from Lorne Gill, Scottish Natural Heritage).

Photo 5. Bassenthwaite Lake in the Lake District, England; example of Group D (photograph from Peter Wakely, English Nature).

Photo 6. Loch of Hundland, mainland Orkney, Scotland; example of Group E (photograph from SNH Orkney Office, Scottish Natural Heritage).

Photo 7. Slapton Ley, Devon, England; example of Group F (photograph from Peter Wakely, English Nature).

Photo 8. The Mere, Ellesmere, Shropshire, England; example of Group G (photograph from Peter Wakely, English Nature).

Photo 9. Mill Loch, Eday, Orkney, Scotland; example of Group H (photograph from SNH Orkney Office, Scottish Natural Heritage).

Photo 10. Malham Tarn, north Yorkshire, England; example of Group I (photograph from George Hinton, English Nature).

Photo 11. Loch of the Stack, Westray, Orkney, Scotland; example of Group J (photograph from SNH Orkney Office, Scottish Natural Heritage).

Chapter 6: Using the Plant Lake Ecotype Index (PLEX)

Eutrophication and acidification are the two major forms of pollution in lakes in Britain. Independent studies have shown that aquatic macrophytes respond to these environmental impacts through changes in species composition and abundance (Farmer 1990; Moss *et al*. 1996; Moss 1998). Ideally, any lake environmental change index should be responsive to these two pressures. The correlation between PLEX and pH and alkalinity has already been demonstrated in Chapter 3 (Figures 3.4, 3.5). In this chapter PLEX is applied to two long-term macrophyte datasets to explore its response to changing scenarios of nutrient input and/or pH. In both cases, the PLEX score was calculated using the submerged and floating taxa recorded in the site and given a PLEX value in Annex B of this report.

Case Study One: Llangorse Lake, Wales (Lake Group I)

Llangorse Lake is a shallow, alkaline, nutrient-rich lake in the Brecon Beacons National Park, Wales (Duigan *et al*., 1999). *Phragmites australis* and *Typha latifiolia* are the dominant components of the extensive beds of marginal emergent vegetation. The diverse range of floating and submerged macrophyte taxa in open water is considered sensitive to a range of environmental impacts, including artificial enrichment, land-use change, power-boating and fishery management. Studies of its environmental history suggest that agricultural intensification in the catchment took place during the early to mid-19th century but the ecology of the lake also appears to have responded to Roman settlement. It was concluded that the lake has been subject to human impact and has been nutrient-rich for a long time (Bennion & Appleby 1999).

The most significant impacts on lake ecology in recent times have been an effluent discharge from sewage treatment works, which entered the lake between 1950s and 1982, and a concurrent intensification of agriculture within the catchment. These pressures led to a significant change in the submerged aquatic flora, which eventually consisted of only a few stands of two species (*Myriophyllum spicatum* and *Potamogeton crispus*) in 1982 (Wade 1999). Following the diversion of the effluent around 1992, there was a significant recovery of the submerged flora in terms of species diversity and abundance. The recovered flora was comparable to the pre-sewage enrichment composition.

As an example of the use of PLEX scores to monitor environmental changes within a lake, we use aquatic macrophyte occurrence data collected at Llangorse Lake over four decades, from 1960 to 1998 (Wade 1999). Figure 1 in Wade (1999) lists the taxa recorded at the lake for each survey year. Following the procedure outlined in Chapter 3, the site PLEX score for each year was calculated by taking the mean of the PLEX scores for each taxon recorded in that year (Table 6.1). These scores were then plotted against year in Figure 6.1.

Table 6.1. Calculating a time series of mean site PLEX scores for Llangorse Lake.

Taxon*	Ced	Chara	Ec	En	Msp	Pbe	Pcr	Pgr	Plu	Ppec	Pper	Ppu	Rc	Zan	Site
PLEX	8.85	7.69	7.95	7.95	8.85	7.69	7.95	7.31	7.88	8.85	7.69	7.95	8.85	8.85	PLEX
pre1960		7.69	7.95		8.85				7.88	8.85	7.69		8.85		8.25
1961	8.85		7.95		8.85		7.95	7.31	7.88		7.69		8.85		8.17
1964			7.95		8.85	7.69			7.88		7.69			8.85	8.15
1969	8.85	7.69			8.85	7.69				8.85	7.69	7.95			8.22
1972		7.69	7.95		8.85		7.95		7.88	8.85	7.69	7.95	8.85	8.85	8.25
1973			7.95		8.85		7.95		7.88	8.85	7.69	7.95	8.85	8.85	8.31
1977		7.69											8.85		8.27
1978													8.85		8.85
1979													8.85		8.85
1980			7.95		8.85		7.95			8.85				8.85	8.49
1981			7.95		8.85									8.85	8.55
1982					8.85		7.95								8.40
1985			7.95		8.85		7.95			8.85			8.85	8.85	8.55
1986			7.95		8.85		7.95			8.85	7.69				8.26
1987		7.69	7.95		8.85					8.85	7.69		8.85		8.31
1989	8.85		7.95		8.35					8.85			8.85		8.67
1990	8.85		7.95		8.35		7.95			8.85	7.69		8.85	8.85	8.48
1991	8.85		7.95		8.35					8.85	7.69				8.44
1992	8.85	7.69	7.95		8.35				7.88	8.85	7.69		8.85		8.33
1995	8.85	7.69	7.95		8.35		7.95		7.88	8.85	7.69	7.95	8.85	8.85	8.30
1998		7.69		7.95	8.85	7.69			7.88	8.85	7.69	7.95	8.85		8.16

See Annex A for full taxon name.

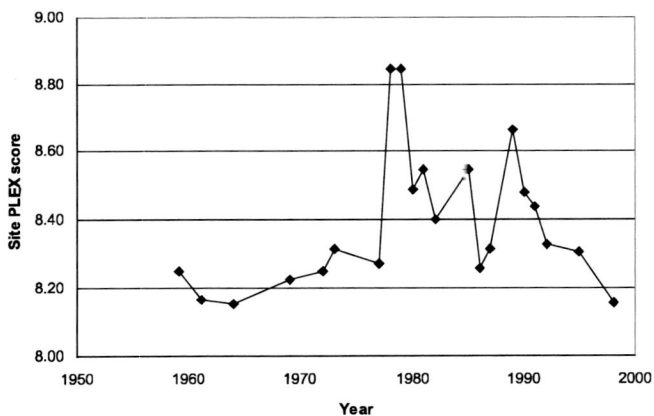

Figure 6.1. PLEX scores through time for Llangorse Lake.

There are evident changes in site PLEX scores, which reflect the recorded changes in the macrophyte composition and abundance at the site. In particular, the maximum PLEX scores are concurrent with the period of most serious enrichment, when the submerged macrophyte communities were dominated by a small number of taxa with high PLEX scores (Table 6.1). There was a trend of gently increasing scores until the late 1970s, then on two occasions scores exceeded 8.8, and since then there have been fluctuations suggestive of a considerable degree of ecosystem instability.

Case Study Two: Loch Chon, Scotland (Lake Group C2)

For over 14 years, the UK Acid Waters Monitoring Network has been collecting biological and chemical data from a series of standing and running water sites scattered throughout England, Scotland, Wales and Northern Ireland (Monteith & Evans 2000; Shilland *et al.* 2002). In the lakes, relative plant taxa abundance has been recorded on a scale similar to DAFOR, following transects, shoreline surveys and deep-water grapnel trawls.

Loch Chon is located in an afforested catchment in central Scotland. It has a relatively diverse aquatic flora for a moderately acid lake with sheltered habitats (Monteith & Evans 2000). *Littorella uniflora* and *Lobelia dortmanna* dominate in the shallow littoral zone, while *Isoetes lacustris*, *Myriophyllum alterniflorum* and *Juncus bulbosus* var. *fluitans* are abundant in deeper water offshore. *Potamogeton berchtoldii*, *Subularia aquatica* and *Elatine hexandra* were recorded from the mid 1990s onwards.

Palaeoenvironmental studies have shown that Loch Chon has undergone dramatic acidification over the last 150 years, with a recent accelerated rate attributed to the afforestation in the catchment (Kreiser *et al.* 1990). Agricultural activities at one end of the loch may have caused mild nutrient enrichment. It is considered possible that the appearance of *P. berchtoldii*, *S. aquatica* and *E. hexandra* (Table 6.2) may have been a response to recent amelioration in acidity, but it could also be a localised response to slight enrichment from agriculture

(Monteith & Evans 2000). The time-series biological data were suggestive of "improved" environmental conditions, but further monitoring is required to resolve whether this was a response to decreasing atmospheric emissions, a sign of artificial enrichment or an indication of a climatic cycle.

The step-like response in the PLEX scores, evident in Figure 6.2, is concurrent with the arrival of the three new species at the site and these elevated values have been maintained.

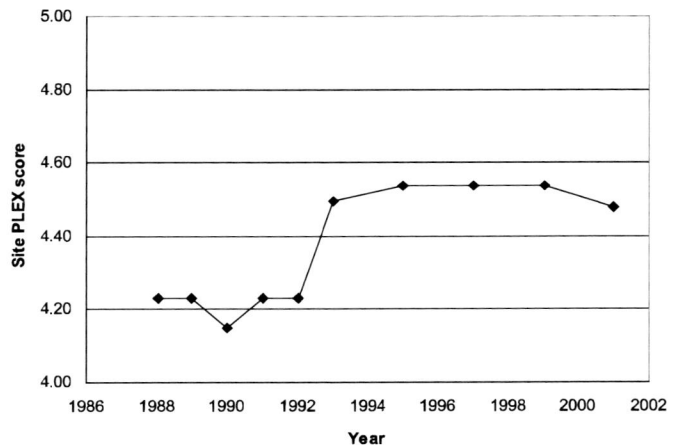

Figure 6.2. PLEX scores through time for Loch Chon.

Table 6.2. Calculating a time series of mean site PLEX scores for Loch Chon.

Taxon*	Pn	Ppol	Spa	Fsq	Isl	Lob	Mal	Ut	Jba	Spa	Lita	Glf	Nul	Na	Pbe	Ela	Sub	Chara	Site
PLEX	4.23	3.08	4.23	5.38	4.23	3.08	4.23	3.08	3.08	4.23	4.23	6.54	6.92	3.08	7.69	5.38	4.23	7.69	PLEX
1988			4.23	5.38	4.23	3.08	4.23	3.08	3.08	4.23	4.23	6.54	6.92	3.08					**4.36**
1989			4.23	5.38	4.23	3.08	4.23	3.08	3.08	4.23	4.23	6.54	6.92	3.08					**4.36**
1990	4.23	3.08	4.23	5.38	4.23	3.08	4.23	3.08	3.08	4.23	4.23	6.54	6.92	3.08					**4.26**
1991			4.23	5.38	4.23	3.08	4.23	3.08	3.08	4.23	4.23	6.54	6.92	3.08					**4.36**
1992			4.23	5.38	4.23	3.08	4.23	3.08	3.08	4.23	4.23	6.54	6.92	3.08					**4.36**
1993			4.23	5.38	4.23	3.08	4.23	3.08	3.08	4.23	4.23	6.54	6.92	3.08	7.69				**4.62**
1995			4.23	5.38	4.23	3.08	4.23	3.08	3.08	4.23	4.23	6.54	6.92	3.08	7.69	5.38	4.23		**4.64**
1997			4.23	5.38	4.23	3.08	4.23	3.08	3.08	4.23	4.23	6.54	6.92	3.08	7.69	5.38	4.23		**4.64**
1999			4.23	5.38	4.23	3.08	4.23	3.08	3.08	4.23	4.23	6.54	6.92	3.08		5.38	4.23	7.69	**4.64**
2001			4.23	5.38	4.23	3.08	4.23	3.08	3.08	4.23	4.23	6.54	6.92	3.08			4.23	7.69	**4.59**

* See Annex A for full taxon name.

Conclusions

These two case studies show that a change in taxonomic composition of the submerged and floating plant community is reflected by PLEX. Palmer (1992) suggested that a gross change in species composition would result in the site keying-out in a different lake group. Manipulation and re-analysis of the dataset used in this study has confirmed this hypothesis. However, in these two case studies, the shifts in PLEX values are within the range expected of group C1 and group I lakes (see Figure 3.3).

Palmer (1992) emphasised the importance of using consistent survey methodology and recording during successive surveys. The use of an index such as PLEX requires adherence to this advice. PLEX is a product of the appearance or disappearance of plant taxa, and this could be regarded as a limitation of the scheme. This makes it vulnerable to misinterpretation where a cyclical pattern of plant succession may be occurring. The further analysis of long-term datasets, such as those collected as part of the UK Acid Waters Monitoring Network, offers the potential future opportunity of producing an index capable of reflecting changes in relative abundance of the aquatic plant taxa.

In the case of Loch Chon, it is evident that PLEX can provide a useful, simple, semi-quantitative measure of environmental change, but the cause of the change is not always clearly evident. However, on the basis of these two case study trials it appears that PLEX can be used as part of an environmental assessment and in trend analysis. If data are available for different parts of a water body it may be possible to distinguish local effects. PLEX may be applicable as part of the Common Standard Monitoring Scheme for conservation site condition assessment, which is currently under development by JNCC. It is worth considering the use of PLEX as a measure of taxonomic composition (but not abundance) under the WFD. It is evident that PLEX is clearly linked to the key variables alkalinity and pH (Figures 3.4, 3.5) which relate to nutrient status, and it provides a comparable measure of macrophyte status. However interpretation of changes in PLEX scores needs to be informed by background knowledge of the lake system and the likely responses to environmental pressures. For example, together these two cases studies show that an increase in PLEX scores may indicate a deterioration or improvement in environmental conditions.

48

Chapter 7: Discussion

The lake resource

The glacial history of Britain has left a rich legacy of lakes throughout England, Scotland and Wales. This natural resource has been supplemented by lakes produced by other geomorphological processes (e.g. coastal lagoon formation) and by basins of artificial origin. The latter were often built to enhance the landscape, or produced as a by-product of industry (e.g. peat excavations, clay pits, quarries). Distinctive natural lake groupings, such as the English Lake District or the Shropshire-Cheshire Meres, have long attracted the attention of limnologists (e.g. Pearsall 1920a,b; Reynolds 1979). However, the standing waters of Britain are spread throughout the country (Figure 7.1), and recently they have been grouped into a series of hydro-morphological types, in response to a requirement of the WFD.

This WFD typology divides lakes potentially into 12 types, according to the base status of their drainage water (or catchment geology) and their mean depth (Phillips, unpublished). Unfortunately, there is no lake biological dataset that comprehensively covers this variety of physical types, but the JNCC macrophyte dataset has been used to check whether the typology has any biological relevance. A reasonable correspondence was found between lakes Groups C2, D, E, F, G and I and base status. This is not surprising, as alkalinity ranges help to define the lake groups described in this report (Figure 4.5). However, it is evident that further lake survey work, especially in England and Wales, is probably required to collect biological data covering the range of hydro-morphological types (compare Figure 3.1 with Figure 7.1). At the moment, the JNCC lake macrophyte dataset is probably one of the widest ranging sources of ecological data for lakes in Britain.

Britain's standing waters have been colonised by a diversity of aquatic plants, occurring as distinct assemblages influenced by environmental factors such as altitude and area (Jones *et al*. 2003). Palmer *et al*. (1992) included a comparison between the botanical classification of lakes and earlier attempts by Spence (1964) to classify aquatic macrophyte communities, and the National Vegetation Classification (Rodwell *et al*. 1995), which was still in preparation at the time.

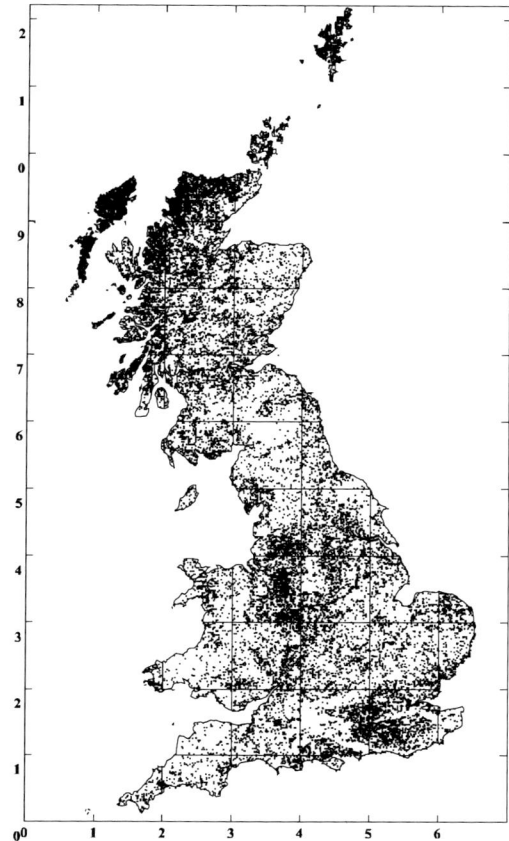

Figure 7.1. Distribution of standing waters in Britain. Taken from 1:50,000 Ordnance Survey mapping for larger water bodies (> 1 ha surface area - or 100m x 100m) in Britain produced from the GB Lakes Inventory (http://ecrc.geog.ucl.ac.uk/gblakes/; Hughes et al. 2004).

The revised classification of Britain's lakes presented here attempts to move the emphasis from an almost exclusive reliance on lists of plant taxa towards a more holistic consideration of lake environments, as reflected by their submerged and floating vegetation. The results demonstrate these lakes are a very important biodiversity resource, responsive to a range of physical and chemical variables, as exemplified by lake Groups A to J.

49

More than 60% of the surveyed lakes in Britain are confined to Groups A-C2, which tend to have largely a north-western distribution with low alkalinity, conductivity and pH, often at relatively high altitude (Figure 7.2). This high proportion is reflective of the intensive survey efforts carried out in these areas (see below) but it has yet to be established as a natural environmental bias in the British lake resource. Although Groups F-I, which are relatively lowland and have high pH, conductivity and alkalinity, make up only 19% of the surveyed sites, they do have distinctive plant assemblages, such as the water-lily dominated lakes of Group F. Future surveys should be directed at potential sites in this series to provide a means of further characterisation and ecological understanding. In particular, the lowland lakes of Group H, with their low plant diversity, seem to have little correspondence with previous attempts to classify British aquatic flora. The same proportion of lakes represented within Groups D and E (16%) is characterised by their relatively high taxon richness. It is suggested that representation of the smaller groups could be substantially increased if attempts were made to supplement the dataset with sites covering the range of WFD lake types. This strategy could also lead to further significant refinements of the lake groups.

Analysis and classification schemes

Table 7.1 shows the distribution of 3445 sites amongst the Lake Groups of the new classification and the Site Types of the previous one (Palmer *et al.*, 1992). The 1992 classification covered only 1118 sites, but the rest of the lakes in the present JNCC database were allotted a Site Type using the TWINSPAN key produced during the first classification exercise.

Few of the Lake Groups have direct Site Type equivalents, but despite considerable overlap there is an obvious relationship between the classifications. The chief anomaly is the distribution of waters in Site Type 8, identified as a nutrient-rich assemblage in the original classification. As expected, Type 8 sites occur in hard water ecotypes in the new classification, but there is also a surprisingly large number allotted to the acidic Lake Group A. This is explained by the preponderance in the present JNCC database of species-poor *Sphagnum* pools that lack the other indicator species for Site Type 1. Therefore, when the original TWINSPAN key is used with the enlarged dataset, the first division leads in the wrong direction. This emphasises the importance of referring to the constancy table, as well as the key, when classifying a site.

The nearest true equivalent of Lake Type A (Ecotype V) is Site Type 1. The soft waters of Ecotype W encompass other examples of Site Type 1 and a large majority of site in Types 2 and 3, which are the upland lakes of the first classification. Ecotype X is populated by many of the Type 4 and 5A sites originally regarded as mesotrophic waters, as well as some Type 3 lakes. Many of the sites in Lake Group F are recognisable as the 'water-lily' lakes of Type 9. The rest of Ecotype Y consists largely of the rich lowland waters of Types 8 and 10A. Ecotype Z includes many of the lakes in Types 7 and 10B, both of which have a high constancy of *Chara* species. Lake Group J is closest to the brackish Site Type 6, but no direct comparison is possible because the later analysis excluded 13 obviously saline sites that formed the majority of the Type 6 grouping.

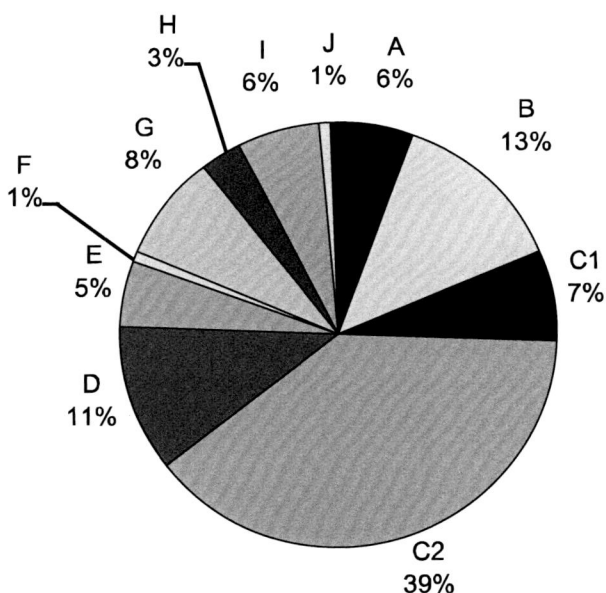

Figure 7.2. Pie chart showing percentage distribution of lakes surveyed within groups.

Table 7.1. The relationship between Lake Groups in the new classification and Site Types in the previous classification: numbers of sites within each class.

Site Types	V	W			X		Y			Z		Total
	A	B	C1	C2	D	E	F	G	H	I	J	no. sites
1	114	107	38	1	4	0	0	0	0	0	0	264
2	2	239	102	640	14	15	2	4	1	1	0	1020
3	0	28	97	656	219	60	0	2	3	2	0	1067
4	0	5	5	15	33	72	0	4	0	20	0	154
5A	0	1	0	2	42	27	0	11	0	4	0	87
5B	0	7	1	2	5	0	0	0	0	0	1	16
6	0	0	0	0	0	0	0	0	0	0	2	2
7	0	0	0	0	2	4	0	5	14	67	5	97
8	102	17	13	3	35	1	20	120	81	9	14	415
9	4	22	0	0	10	0	26	31	0	0	0	93
10A	0	0	0	0	4	1	0	81	0	50	2	138
10B	0	0	0	0	2	6	0	22	2	50	10	92
Total no. sites	222	426	256	1319	370	186	48	280	101	203	34	3445

Figures in **bold**: 40% or more of the total number of sites in the relevant Lake Group (new classification)
Figures in ***bold italic***: 40% or more of the total number of sites in the relevant Site Type (previous classification)
Figures in **bold underlined**: 40% or more of the total number of sites in both the relevant Lake Group and the Site Type
NB. 2 of 3447 sites in the JNCC database were not classified under the old system.

	A	B	C1	C2	D	E	F	G	H	I	J
Natural dystrophic lakes and ponds	■	■	■	■							
Oligotrophic waters containing very few minerals of sandy plains (*Littorelletalia uniflorae*)		■		■							
Oligotrophic to mesotrophic standing waters with vegetation of the *Littorelletea uniflorae* and/or of the *Isoëto-Nanojuncetea*	■	■	■	■	■						
Hard oligo-mesotrophic waters with benthic vegetation of *Chara* spp.	■		■		■	■	■		■		
Natural eutrophic lakes with *Magnopotamion* or *Hydrocharition*-type vegetation					■		■		■		
Coastal lagoons					■						■

Figure 7.3. An illustration of the relationship between the lake groups described in this report and those protected under the Habitats Directive. The closest correspondence is found between groups shaded dark grey but equally important representatives or regional variants may occur in groups shaded dark grey.

Palmer (2001) made a number of recommendations for the treatment of macro-algae and other lower plant taxa in any future analysis. In particular, she recommended that the charophytes, and where possible bryophytes, should be included at species level, and a number of other macro-algae, such as *Enteromorpha*, should be included at generic level. These improvements to the dataset were incorporated as far as possible. For example, a preliminary analysis was attempted with the charophytes at the species level. However, because many of the records were at generic level the use of species was finally abandoned. In the final analysis used, *Enteromorpha* and fucoid algae did occur with high frequency in Group J. *Fontinalis antipyretica* and/or *Sphagnum* spp. showed a high level of constancy in Groups A, B, C1, C2, D, E, F, H and I. There remains scope to extend the range of macroalgae and bryophytes recorded as part of future lake surveys.

Palmer (2001) also recommended that the use of quantitative records should be investigated for monitoring purposes. To substitute for presence/absence data, the pseudospecies cut levels (1, 2, 3, 4 and 5) for the present TWINSPAN analysis were chosen to match the DAFOR scale. The next step will be to develop an environmental change index based on quantitative data. Like the earlier TRS scheme, PLEX is still limited to the use of presence/absence data, unlike the Mean Trophic Rank (MTR) scheme used for rivers which incorporates a 'species cover value' (Holmes 1995; Dawson *et al.* 1999).

No classification system generated by statistical analysis is 100% satisfactory. Any ecological classification generated is an artificial partitioning of a continuum or multi-dimensional space. This is especially true of TWINSPAN or CCA, but it is reassuring to see that correlations are possible between independently derived classifications. For example, this report demonstrates that it is possible to use the lake classification presented as a means of pinpointing sites which may qualify for selection as SACs under the Habitats Directive (Figure 7.3). It is also possible to make links with aquatic plant communities described by the NVC. Efforts are now being made to combine the WFD typology and the lake groups to facilitate the allocation of standing waters to the published Biodiversity Action Plans for mesotrophic lakes and eutrophic standing waters.

Regional studies

On a regional level, Scotland continues to have the most extensive data holdings for plants in standing waters in Britain. The Loch Survey was advanced as a corporate priority by the Nature Conservancy Council for Scotland, and subsequently Scottish Natural Heritage. Survey sites were chosen using a matrix based on area and altitude classes, to which a set of criteria was applied to identify priority sites. The basic aim was to carry out a stratified random survey, but a bias was introduced towards sites on base-rich geology (as they were thought likely to support a high diversity of plants), and all sites > 1 ha in size lying in existing or potential Sites of Special Scientific Interest. However, the supporting environmental dataset (especially water chemistry) for Scotland has limited application, partly due to the one-off nature of most of the surveys. In England, the historic survey effort also appears concentrated on existing or potential SSSIs, and it is understandably focused on areas with the highest concentrations of natural water bodies, such as the Lake District and the Shropshire-Cheshire Meres. In Wales, the lake survey effort had the objective of surveying a representative series of sites within the region, while responding to management information needs at key sites. The distribution of lake groups between countries (Figure 4.7) needs to be interpreted with caution as it is a product of these regional survey strategies. However, at the moment the composition of lake groups in Wales reflect those of Scotland, with C2 lakes recorded at highest frequency. In contrast, representatives of Groups F and G occur with highest frequency in England. At the moment this summary provides a snapshot of the recorded variety of lakes in each country and therefore helps to identify unrecorded or potentially regional rarities. The exact representation of lake groups in particular countries will become clear as further surveys are carried out. For example, a recent survey of lakes in the Migneint-Arenig-Dduallt cSAC has identified Llyn Tryweryn as a Group A lake in Wales (Carvalho *et al.* 2003).

The Welsh lake dataset collected at 31 lakes over the period 1993-1997 included a wide range of physical and chemical data and quantified assemblages of the following biological groups - epilithic diatoms, surface sediment diatoms, aquatic macrophytes, littoral Cladocera (zooplankton), open water zooplankton and littoral macroinvertebrates. Statistical analysis revealed a single, dominant and very wide environmental gradient from low to high pH, alkalinity, conductivity, major ion and phosphorus conditions (Allott & Monteith 1999). Analysis of the variation within the individual biological groups demonstrated the dominant role of this primary environmental gradient in determining species assemblages. Comparison of the integrated biological TWINSPAN classification with classifications generated for the individual biological groups showed that the integrated scheme corresponded most closely with the aquatic macrophyte TWINSPAN scheme. It was suggested that this resulted from the fact that macrophytes respond to both variations in water chemistry and substrate, and have a key role in habitat availability for other biological groups in lakes. A key conclusion of the Welsh study was that biological variation within all the individual groups studied relates most strongly to the primary environmental gradient, and it can be effectively represented by a small number of environmental variables, especially pH, soluble reactive phosphorus, chlorophyll a, conductivity, total phosphorus and alkalinity. On the basis of this study, it was recommended that a subset of sites in England, Scotland and Wales should be surveyed with the objective of confirming the key environmental factors that influence the composition of aquatic plant assemblages and making links with the other biological elements required by the WFD. The analysis of this type of integrated dataset would also address the concern that a system for classifying lakes according to their macrophyte communities may not provide a reliable means of selecting a representative range of sites for other biological groups e.g. invertebrates (Duigan & Kovach 1994).

It would be valuable to incorporate data from the Northern Ireland Lake Survey (Wolfe-Murphy *et al.* 1992) and lake surveys available from the Republic of Ireland into any further data analysis. However, the findings reported here are similar to those reported following a separate analysis of species and environmental relationships of aquatic macrophytes in lakes in Northern Ireland (Heegaard *et al.* 2001). In particular, they found that the most influential environmental variables were related to local-scale water chemistry, which was highly correlated with altitude because hardwater, nutrient-rich lakes are restricted to the lowlands. They also inferred that this local-scale variation was a strong controlling factor on species composition in a lake, leading to the conclusion that the occurrence of a species in a lake is predominantly controlled by the catchment land-use, especially fertiliser use and farming. This report has

shown that a high level of local-scale variation in lake groups is also visible in the areas of Scotland. For example, with the exception of F, all lake groups are represented on the relatively small land area of Orkney.

The current dataset could also be used to explore the environmental requirements of plant species, which has the potential to contribute to conservation site management and to predict species response to environmental change. The relationships between the environmental variables in the dataset and taxon richness and PLEX are presented in Annex C. In particular, the dataset could be used to explore further the relationships between area, altitude and aquatic plant diversity which have been described from a Cumbrian dataset (Jones *et al.* 2003), but it is already interesting to note that certain lake groups have distinctive altitudinal ranges, predominance of large or small water bodies and/or differences in taxon richness (see chapters 4 and 5).

Future developments

Palmer *et al.* (1992) concluded by predicting "a dual analysis, involving both site and vegetation classification, will be employed to carry out site selection and conservation at a European level." This approach is clearly visible in the requirements of the WFD. There is a requirement to type a lake based on its physical characteristics, define its reference or pristine conditions, and then reach a judgement on its ecological quality, using data from a variety of biological elements, including macrophytes.

As part of this process, the revised classification presented in this report is an attempt to make some of the required links between environmental variables and aquatic vegetation. Further research is needed to establish more precisely how environmental variables influence aquatic plant distribution within and between lakes. For a lake at a precise location, defined altitude, known geology, water chemistry and depth, what assemblage of aquatic plants would be expected? This type of investigation is now being advanced in Britain as part of the supporting research for the WFD. It is important to remember that the classification scheme presented in this report is equivalent to an environmental snapshot of the condition of a lake at the time of survey, that the data used to draw up the classification were collected over a long period of time, and that the lakes surveyed ranged from examples in a pristine or near-pristine condition to sites that have been heavily degraded. Palmer (2001) recommended that any future index of environmental change should be reference based, to enable the degree of departure from pristine water quality to be taken into account. A growing list of lakes considered to be at reference condition in Britain is being assembled, but it remains to be seen if a sufficient number can be identified to form the basis of a future classification scheme.

Finally, the publication of this report brings a requirement to update the JNCC macrophyte database, using the revised lake groupings, and the revision of SSSI selection guidelines (Nature Conservancy Council 1989), which are based on the earlier botanical classification. The revised classification scheme will continue to provide an essential element in the process of site selection for conservation, but it needs to be supplemented by environmental data. Newly surveyed lakes can be classified using the key presented (Table 4.2). Together with PLEX, these practical applications will contribute to information required for conservation site selection and management.

Chapter 8: Acknowledgements

The authors are grateful to Philip Boon, chair of the Joint Nature Conservation Committee Freshwater Lead Coordination Network for his support of this project. Other colleagues in the conservation agencies provided assistance with background information, comments, data and photographs, and report production, especially Mary Hennessy (Scottish Natural Heritage), Alison Lee and Susan Watt (Joint Nature Conservation Committee), Kath Lees (Scottish Natural Heritage) and Stewart Clarke (English Nature).

Geoff Phillips (Environment Agency), Ian Fozzard (Scottish Environment Protection Agency) and Ian Strachan (Joint Nature Conservation Committee) provided constructive criticism of the results and the report.

The authors are indebted to the various teams of surveyors that carried out the fieldwork in all types of weather conditions, especially the members of the Scottish Loch Survey. Laurence Carvalho and Don Monteith (Environmental Change Research Centre, University College London) carried out the recent macrophyte surveys in England and Wales. Chris Newbold, Margaret Palmer and Liz Charter collected much of the earlier data in these two countries.

Don Monteith also allowed used of the Acid Waters Monitoring Data from Loch Chon. Stephen Maberly and Julie Parker (Centre for Ecology and Hydrology, Lancaster) made corrections to the alkalinity dataset. ENIS Ltd. assisted with dataset compilation.

Chapter 9: References

Allott, T.E.H., & Monteith, D.T. 1999. *Classification of Lakes in Wales for Conservation using Integrated Biological Data.* Countryside Council for Wales Contract Science Report No. 314. Bangor.

Allott, T.E.H., Monteith, D.T., Duigan, C.A., Bennion, H., & Birks, H.J.B. 2001. *Conservation Classification of the lakes in Wales, with implications for the EU Water Framework Directive.* Countryside Council for Wales Contract Science Report No. 426. Bangor.

Bennion, H., & Appleby, P. 1999. An assessment of recent environmental change in Llangorse Lake using palaeolimnology. *Aquatic Conservation: Marine and Freshwater Ecosystems, 9:* 361-375.

Carvalho, L., Reynolds, B., Lyle, A., Norris, D., & Brittain, A. 2003. *Strategic CCW Conservation Lake Survey: A survey of lakes in the Migneint-Arenig-Dduallt pSSSI/SAC/SPA, Wales, 2002-3: Final Report.* Countryside Council for Wales Contract Science Report No. 592. Bangor.

CEN 2006. *Water Quality - Guidance standard for the surveying of macrophytes in lakes.* pr EN 15460

Dawson, F.H., Newman, J.R., Gravelle, M.J., Rouen, K.J., & Henville, P. 1999. *Assessment of the Trophic Status of Rivers Using Macrophytes. Evaluation of the Mean Trophic Rank.* Environment Agency Research and Development Technical Report E39. Bristol.

Duigan, C.A. & Kovach, W.L. 1994. Relationships between littoral microcrustacea and aquatic macrophyte communities on the Isle of Skye (Scotland), with implications for the conservation of standing waters. *Aquatic Conservation: Marine and Freshwater Ecosystems, 4:* 307-331.

Duigan, C., Kovach, W. & Palmer, M. 2002. A botanical classification of British lakes: preliminary results. *British Ecological Society Winter Meeting, 18-20 December, 2002. Programme and abstracts.*

Duigan, C.A., Reid, S., Monteith, D.T., Bennion, H., Seda, J.M., & Hutchinson, J. 1999. The past, present and future of Llangorse Lake - a shallow nutrient-rich lake in the Brecon Beacons National Park, Wales, UK. *Aquatic Conservation: Marine and Freshwater Ecosystems, 9:* 329-341.

Duigan, C.A., & Phillips, G. 2002. Aquatic macrophyte and the Water Framework Directive: ongoing developments for British lakes. *TemaNord 2002:* 566, 54.

Farmer, A.M. 1990. The effects of lake acidification on aquatic macrophytes - a review. *Environmental Pollution, 65:* 219-240.

Foster, D., Wood, A. & Griffiths, M. 2001. The EC Water Framework Directive and its implications for the Environment Agency. *Freshwater Forum, 16:* 4-28.

Heegaard, E., Birks, H.H., Gibson, C.E., Smith, S.J., & Wolfe-Murphy, S. 2001. Species-environmental relationships of aquatic macrophytes in Northern Ireland. *Aquatic Botany, 70:* 175-223.

Hill, M.O. 1979. *TWINSPAN - A FORTRAN program for arranging multivariate data in an ordered two-way table by classification of the individuals and attributes.* Cornell University, Ithaca, New York.

Holmes, N.T.H. 1995. *Macrophytes for Water and Other River Quality Assessments. A report to the National Rivers Authority.* National Rivers Authority, Anglian Region, Peterborough.

Hughes, M. Hornby, D.D., Bennion, H., Kernan, M., Hilton, J., Phillips, G. & Thomas, R. 2004. The development of a GIS-based inventory of standing waters in Great Britain together with a risk-based prioritisation protocol. *Water, Air and Soil Pollution: Focus 4:* 73-84.

IUCN Species Survival Commission. 2000. *IUCN Red List Categories and Criteria. Version 3.1. As Approved by the 51st. Meeting of the IUCN Council, Gland, Switzerland.* IUCN - The World Conservation Union, Gland, Switzerland.

Jones, J.I., Li, W., & Maberly, S.C. 2003. Area, altitude and aquatic plant diversity. *Ecography, 26:* 411-420.

Kovach, W.L. 2001. *MVSP - A MultiVariate Statistical Package, ver. 3.12.* Kovach Computing Services, Pentraeth, Wales, U.K.

Kreiser, A.M., Appleby, P.G., Natkanski, J., Rippey, B., & Battarbee, R.W. 1990. Afforestation and lake acidification: a comparison of four sites in Scotland. *Philosophical Transactions of the Royal Society of London, Series B, 327:* 377-383.

Lassière, O. 1998. *Botanical Survey of Scottish Freshwater Lochs: Methodology.* Scottish Natural Heritage Draft Report.

Monteith, D.T., & Evans, C.D. (editors) 2000. *UK Acid Waters Monitoring Network: 10 Year Report. Analysis and Interpretation of Results, April 1988-March 1998.* ENSIS Ltd, London.

Moss, B. 1998. *Ecology of Fresh Waters. Man and Medium, Past to Future.* Blackwell Science. Oxford.

Moss, B., Madgwick, J., & Phillips, G. 1996. *A guide to the restoration of nutrient-enriched shallow lakes.* Environment Agency, Broads Authority & European Union Life Programme.

Nature Conservancy Council 1989 (rev. ed. Joint Nature Conservation Committee 1998). *Guidelines for selection of biological SSSIs.* Peterborough.

Oksanen, J., & Minchin, P. R. 1997. Instability of ordination results under changes in input data order: explanations and remedies. *Journal of Vegetation Science, 8:* 447-454.

Palmer, M. 1992. *A botanical classification of standing waters in Great Britain and a method for the use of macrophyte flora in assessing changes in water quality incorporating a reworking of data 1992.* Peterborough, Joint Nature Conservation Committee. (Research and Survey in Nature Conservation, No. 19.)

Palmer, M.A. 2001. An approach to the use of macrophytes for monitoring standing waters. *Freshwater Forum, 16:* 82-90.

Palmer, M. In prep. *LACON: Lake Assessment for Conservation.* Scottish Natural Heritage Commissioned Report.

Palmer, M.A., Bell, S.A., & Butterfield, I. 1992. A botanical classification of standing waters in Britain: applications for conservation and monitoring. *Aquatic Conservation: Marine and Freshwater Ecosystems, 2:* 125-143.

Parr, T.W., Monteith, D.T., & Gibson, M. 1999. Aquatic Macrophytes. *In: The United Kingdom Environmental Change Network - Protocols for Standard Measurements at Freshwater Sites,* ed. by J.M. Sykes, A.M.J. Lane and D.G. George, 66-79. Natural Environment Research Council. Abbotts Ripton, Huntington.

Paton, J.A. 1999. *The Liverwort Flora of the British Isles.* Harley Books. Martins, Great Horkesley, Colchester, Essex, England.

Pearsall, W.H. 1920a. The aquatic vegetation of the English Lakes. *Journal of Ecology, 8:* 163-201.

Pearsall, W.H. 1920b. The development of vegetation in the English Lakes, considered in relation to the general evolution of glacial lakes and rock-basins. *Proceedings of the Royal Society B, 92:* 259-284.

Pokorný J. & Květ, J. 2004. Aquatic Plants and Lake Ecosystems. In: *The Lakes Handbook. Limnology and Limnetic Ecology,* ed. By P.E. O'Sullivan and C.S. Reynolds, 309-340. Blackwell Publishing. Oxford, UK.

Pollard, P. & Huxham, M. 1998. The European Water Framework Directive: a new era in the management of aquatic ecosystem health? *Aquatic Conservation: Marine and Freshwater Ecosystems, 8:* 773-792.

Preston, C.D. 1995. *Pondweeds of Great Britain and Ireland.* Botanical Society of the British Isles. London.

Preston, C.D., & Croft, J.M. 1997. *Aquatic Plants in Britain and Ireland.* Harley Books. Martins, Great Horkesley, Colchester, Essex, England.

Preston, C.D., Pearman, D.A., & Dines, T.D. 2002. *New atlas of the British and Irish Flora - An Atlas of Vascular Plants of Britain, Ireland, the Isle of Man and the Channel Islands.* Oxford University Press. Oxford.

Reynolds, C. S. 1979. The limnology of the eutrophic meres of the Shropshire-Cheshire plain: a review. *Field Studies, 5:* 93-173.

Rodwell, J.S. (editor), Pigott, C.D., Ratcliffe, D.A., Malloch, A.J.C., Birks, H.J.B., Proctor, M.C.F., Shimwell, D.W., Huntley, J.P., Radford, E., Wigginton, M.J. & Wilkins, P. 1995. *Aquatic Communities, Swamps and Tall-herb Fens.* British Plant Communities. Volume 4. Cambridge University Press. Cambridge.

Scheffer, M. 1998. *Ecology of Shallow Lakes.* Chapman & Hall. London.

Shilland, E.M., Monteith, D.T., Smith, J., & Beaumont, W.R.C. (editors) 2002. *The United Kingdom Acid waters Monitoring Network Data Report for 2001-2002 (Year 14).* Report to the Department for Environment, Food and Rural Affairs and the Department of the Environment Northern Ireland. ENSIS Ltd, London.

Spence, D.H.N. 1964. The macrophytic vegetation of lochs, swamps and associated fens. *In : The Vegetation of Scotland,* ed. by J.H. Burnett, 306-425. Oliver and Boyd Ltd., Edinburgh.

Stace, C. 1991. *New Flora of the British Isles.* Cambridge University Press. Cambridge.

Stewart, N.F. & Church, J.M. 1992. *Red Data Books of Britain and Ireland: Stoneworts.* Joint Nature Conservation Committee, Peterborough.

Stewart, A., Pearman, D.A. & Preston, C.D. 1994. *Scarce Plants in Britain.* Joint Nature Conservation Committee, Peterborough.

ter Braak, C.J.F. 1986. Canonical correspondence analysis: A new eigenvector technique for multivariate direct gradient analysis. *Ecology, 67:* 1167-1179.

Wade, P.M. 1999. The impact of human activity on the aquatic macroflora of Llangorse Lake, South Wales. *Aquatic Conservation: Marine and Freshwater Ecosystems, 9:* 441-459.

Wetzel, R.G. 2001. *Limnology, Lake and River Ecosystems.* Academic Press.

Wigginton, M.J. (ed.) 1999. *British Red Data Books 1. Vascular Plants,* 3rd. edition. Joint Nature Conservation Committee, Peterborough.

Willby, N.J., Pygott, J.R. & Eaton, J.W. 2001. Inter-relationships between standing crop, biodiversity and trait attributes of hydrophytic vegetation in artificial waterways. *Freshwater Biology, 46:* 883-902.

Wolfe-Murphy, S.A., Lawrie, E.A., Smith, S.J. & Gibson, C.E. 1991. *Survey methodologies: data collection techniques.* A report by the Northern Ireland Lakes Survey, Department of the Environment (Northern Ireland), Belfast.

Wolfe-Murphy, S.A., Lawrie, E.A., Smith, S.J. & Gibson, C.E. 1992. The Northern Ireland Lake Survey: Part 3. Lake classification based on aquatic macrophytes. Department of the Environment and Queen's University, Belfast.

Chapter 10: Annexes

Annex A: A listing of submerged and floating macrophyte taxa included in the 1989 TWINSPAN analysis (Palmer *et al.* 1992) and the 2004 re-analysis of the JNCC lake dataset described in this report, with the number of records used. Any subsequent analysis of this dataset should try to incorporate these taxa. Most of the differences between the records used in 1989 and 2004 have arisen from uncertainty over the form (submerged/floating/emergent) of the plants recorded, taxonomic amalgamations, and the inclusion of canals in the first dataset.

Code	Scientific Name	1989	2004	Notes
Ag	*Alisma gramineum*	1	1	One record (EN0303 - Westwood Great Pool) added by MP to reanalysis.
Af	*Azolla filiculoides*	2	1	
Api	*Apium inundatum*	38	186	
Cah/a	*Callitriche hamulata*	152	585	Records merged with *C. hamulata* sens.lat (Cah).
Cao	*Callitriche obtusangula*	15	14	
Cas/a	*Callitriche stagnalis*	206	664	Records merged with *C. stagnalis* sens.lat (Cas).
Ctrunc	*Callitriche truncata*	4	1	One record (EN0073 - Clumber Park Lake) added by MP to re-analysis.
Caa	*Catabrosa aquatica*	1	0	Included by Palmer (1992) but not in 2004 re-analysis.
Ced	*Ceratophyllum demersum*	28	44	
Cersub	*Ceratophyllum submersum*	3	9	
Chara	*Chara* spp.	279	726	Cha (in database); Chara (in spreadsheet) *Chara* only taxon included by Palmer (1992). In the re-analysis the following *Chara* records were merged and the highest DAFOR rating was used - Cha, Chaas, Chaas/s, Chaas/l, Chaco, Chacu, Chagl,Chaglsl, Chahi, Chape, Charu, Chavi, Chavi/a, Chavu, Chavu/h, Chavu/l, Chavu/p, Chavu/v, Chavusst,
Cher	*Callitriche hermaphroditica*	87	168	
Cpla	*Callitriche platycarpa*	23	23	
Crh	*Crassula helmsii*	2	2	
Drepan	*Drepanocladus* spp. and *Drepanocladus fluitans*	8	0	Palmer (1992) used *Drepanocladus* spp. (n = 6) and *D. fluitans* (n =2).
Ec	*Elodea canadensis*	183	316	
Ef	*Eleogiton fluitans*	90	523	= *Scirpus fluitans*
Ela	*Elatine hexandra*	25	75	
Elaca	*Eleocharis acicularis*	21	34	Records for Elaca (aquatic) and Elac used.
Elh	*Elatine hydropiper*	8	15	
En	*Elodea nuttallii*	58	52	
Ent	*Enteromorpha* spp.	42	60	Amalgamated records for *Enteromorpha* spp. (Ent) and *E. intestinalis* (Entint)
Ese	*Eriocaulon aquaticum*	3	48	
Fon	*Fontinalis antipyretica*	262	803	
Fsq	*Fontinalis squamosa*	1	0	
Fuc	*Fucoid algae*	10	11	
Gld	*Glyceria declinata*	11	0	
Glf	*Glyceria fluitans*	289	1052	
Gln	*Glyceria notata*	8	5	
Grd	*Groenlandia densa*	3	4	
Hipa	*Hippuris vulgaris*	176	153	Records for Hipa (aquatic) and Hip used.
Hotpal	*Hottonia palustris*	5	13	
Hmr	*Hydrocharis morsus-ranae*	6	12	

Code	Scientific Name	1989	2004	Notes
Ise	*Isoetes echinospora*	5	50	
Isl	*Isoetes lacustris*	215	832	
Jba	*Juncus bulbosus*	537	2140	Records for Jb and Jba (aquatic) used
Lam	*Lagarosipon major*	0	3	
Lg	*Lemna gibba*	5	12	
Lita	*Littorella uniflora*	565	1838	Records for *Littorella uniflora* (Lit) and aquatic form (Lita) merged
Lm	*Lemna minor*	148	409	
Lmi	*Lemna minuta*	1	1	
Lob	*Lobelia dortmanna*	313	1383	
Lt	*Lemna trisulca*	53	102	
Lun	*Luronium natans*	6	8	
Mal	*Myriophyllum alterniflorum*	453	1552	
Msp	*Myriophyllum spicatum*	183	216	
Myrver	*Myriophyllum verticillatum*	1	3	
Na	*Nymphaea alba*	237	872	
Nf	*Najas flexilis*	11	22	
Nm	*Najas marina*	1	0	
Nit or Nitella	*Nitella* spp.	175	686	*Nitella* only taxon used by Palmer (1992). Merged records for Nit, Nitco, Nitflagg, Nitgr, Nitmu, Nitop, Nittr for re-analysis.
Nul	*Nuphar lutea*	131	214	
Nup	*Nuphar pumila*	28	58	
Nus	*Nuphar lutea* x *pumila*	4	14	*N. intermedia* in Palmer (1992).
Nyp	*Nymphoides peltata*	5	11	
Oa	*Oenanthe aquatica*	13	12	
Oenflu	*Oenanthe fluviatilis*	2	0	
Pal	*Potamogeton alpinus*	65	169	
Pama	*Persicaria amphibia*	216	306	= *Polygonum amphibium*; records for *P amphibia* (Pam) and aquatic form (Pama) merged for reanalysis.
Pbe	*Potamogeton berchtoldii*	190	416	
Pcol	*Potamogeton coloratus*	0	9	
Pcom	*Potamogeton compressus*	4	1	
Pcr	*Potamogeton crispus*	104	194	
Pep	*Potamogeton epihydrus*	0	1	
Pfil	*Potamogeton filiformis*	112	169	
Pfr	*Potamogeton friesii*	30	28	
Pgr	*Potamogeton gramineus*	163	278	
Pgr/pxn	*Potamogeton gramineus* x *nitens*	0	3	
Pil	*Pilularia globulifera*	12	25	Merged records for Pil and Pila (aquatic) for re-analysis.
Plu	*Potamogeton lucens*	41	34	
Pn	*Potamogeton natans*	471	1647	
Pob	*Potamogeton obtusifolius*	85	181	
Ppec	*Potamogeton pectinatus*	199	220	
Pper	*Potamogeton perfoliatus*	242	450	
Ppol	*Potamogeton polygonifolius*	330	1633	
Ppra	*Potamogeton praelongus*	32	88	
Ppu	*Potamogeton pusillus*	142	187	
Pr	*Potamogeton rutilus*	4	8	
Pt	*Potamogeton trichoides*	17	11	

Code	Scientific Name	1980	2004	Notes
Pxg	*Potamogeton natans x polygonifolius*	0	1	
Pxgr	*Potamogeton alpinus x praelongus*	0	1	
Pxn	*Potamogeton gramineus x perfoliatus*	13	127	
Pxsp	*Potamogeton gramineus x natans*	0	4	
Pxz	*Potamogeton gramineus x lucens*	5	27	Included by Palmer (1992) as *P. x zizii*
No code	*Potamogeton x lintonii*	1	0	Included by Palmer (1992); = *P. crispus* x *P. friesii*. Taxon not included in JNCC database.
Ra	*Ranunculus aquatilis* sens. str.	51	66	*Ranunculus aquatilis* included by Palmer (1992)
Rb	*Ranunculus baudotii*	82	97	
Rc	*Ranunculus circinatus*	42	47	
Rf	*Ranunculus fluitans*	1	0	
Rh	*Ranunculus hederaceus*	22	74	
Ricflu	*Riccia fluitans*	5	9	
Ro	*Ranunculus omiophyllus*	14	18	
Rp	*Ranunculus peltatus*	28	75	
Rpse	*Ranunculus penicillatus subspecies pseudofluitans*	1	1	
Rtr	*Ranunculus trichophyllus*	32	55	
Ricnat	*Ricciocarpos natans*	2	0	
Ru	*Ruppia* spp.	21	0	Palmer (1992) included *Ruppia* sp.
Ruc	*Ruppia cirrhosa*	0	4	
Rum	*Ruppia maritima*	0	29	
Spa	*Sparganium angustifolium*	260	1140	
Spema	*Sparganium emersum*	60	155	Spem and Spema (aquatic) records amalgamated for reanalysis.
Sphag	*Sphagnum*	80	1004	*Sphagnum* sp. (submerged) included by Palmer (1992) merged with small number of records for other *Sphagnum* taxa
Spn	*Sparganium natans*	71	239	= *Sparganium minimum*
Spo	*Spirodela polyrhiza*	3	5	= *Lemna polyrhiza*
Stralo	*Stratiotes aloides*	5	0	
Sub	*Subularia aquatica*	85	353	
Um	*Utricularia minor*	59	364	
Uti	*Utricularia intermedia* sens. lat.	83	256	*Utricularia intermedia* agg. included by Palmer (1992)
Uva	*Utricularia vulgaris* sens.lat.	89	187	*Utricularia vulgaris* agg. included by Palmer (1992)
No code	*Vallisneria spiralis*	1	0	Not in JNCC database
Woa	*Wolffia arrhiza*	0	1	
Zan	*Zannichellia palustris*	86	145	

Annex B: Constancy table for standing water site types: submerged and floating plants with number of occurrences and PLEX values.

Code	Taxon	A	B	C1	C2	D	E	F	G	H	I	J	Occ.	PLEX
Ag	*Alisma gramineum*								0.36				1	------
Api	*Apium inundatum*	0.45	3.52		2.43	12.97	24.19		6.41	10.89	7.39	2.86	186	7.50
Af	*Azolla filiculoides*								0.36				1	------
Cah/a	*Callitriche hamulata*	0.45	2.82	8.59	11.90	66.76	17.20	6.25	27.05	13.86	9.85	2.86	585	6.15
Cher	*Callitriche hermaphroditica*				0.38	5.14	23.66		11.39	1.98	32.51		168	7.69
Cao	*Callitriche obtusangula*				0.15			4.17	2.85		0.99		14	------
Cpla	*Callitriche platycarpa*				0.30	0.81		8.33	2.14	2.97	1.48		23	------
Cas/a	*Callitriche stagnalis*	0.90	8.69	8.98	9.48	34.59	29.03	47.92	39.50	74.26	38.42	22.86	664	7.69
Ctrunc	*Callitriche truncata*								0.36				1	------
Ced	*Ceratophyllum demersum*				0.54			4.17	9.25	0.99	6.40		44	8.85
Cersub	*Ceratophyllum submersum*							4.17	2.49				9	------
Chara	*Chara* spp.		10.56		18.57	21.89	77.42	6.25	22.06	1.98	68.97	11.43	726	7.69
Crh	*Crassula helmsii*					0.27	0.54						2	------
Ela	*Elatine hexandra*				2.20	8.11	4.84		1.07		1.97		75	5.38
Elh	*Elatine hydropiper*					1.08			3.20		0.99		15	------
Elaca	*Eleocharis acicularis*					1.08	1.61		6.41		4.43		34	7.95
Ef	*Eleogiton fluitans*	0.45	23.00	2.73	28.13	7.03	9.14	2.08			0.49	2.86	523	3.08
Ec	*Elodea canadensis*	0.45	1.88		1.52	20.81	10.22	20.83	44.84	0.99	26.60		316	7.95
En	*Elodea nuttallii*		0.23		0.15	3.24	1.61	2.08	7.47		5.91		52	7.95
Ent	*Enteromorpha* spp.					0.27	1.08	2.08	7.83		7.39	54.29	60	8.85
Ese	*Eriocaulon aquaticum*	0.90	5.16	0.78	1.67								48	3.08
Fon	*Fontinalis antipyretica*	0.90	3.52	15.23	26.31	61.35	50.00	4.17	12.10	7.92	17.73		803	5.38
Fuc	Fucoid algae					0.27	0.54				0.49	22.86	11	------
Glf	*Glyceria fluitans*	4.05	17.14	11.72	26.99	64.05	46.77	12.50	40.93	70.30	31.03	14.29	1052	6.54
Gln	*Glyceria notata*							4.17	0.36		0.99		5	------
Grd	*Groenlandia densa*								1.42				4	------
Hipa	*Hippuris vulgaris*		5.16	1.95	1.06	8.11	5.38	10.42	8.19	5.94	18.72		153	7.88
Hotpal	*Hottonia palustris*		0.23						4.27				13	------
Hmr	*Hydrocharis morsus-ranae*		0.23						3.91				12	------
Ise	*Isoetes echinospora*	0.45	0.23	1.56	2.05	1.62	5.91						50	5.38
Isl	*Isoetes lacustris*	1.35	0.94	28.13	44.96	32.16	21.51		0.36				832	4.23
Jba	*Juncus bulbosus*	49.55	69.01	87.50	92.57	52.70	44.09		2.85		2.46	2.86	2140	3.08
Lam	*Lagarosipon major*								0.71		0.49		3	------
Lg	*Lemna gibba*								4.27				12	------
Lm	*Lemna minor*	1.35	5.63		0.30	11.08	12.37	60.42	75.44	14.85	28.08	2.86	409	8.85
Lmi	*Lemna minuta*								0.36				1	------
Lt	*Lemna trisulca*		0.47			1.89	2.15	12.50	24.56	0.99	6.40		102	8.85
Lita	*Littorella uniflora*		11.03	46.48	88.40	70.00	88.71	2.08	8.54		27.59	2.86	1838	4.23
Lob	*Lobelia dortmanna*	0.90	16.90	37.89	83.55	20.81	17.74						1383	3.08
Lun	*Luronium natans*				0.23	1.35							8	------
Mal	*Myriophyllum alterniflorum*	0.90	14.08	21.48	77.03	62.70	83.87		4.98		8.37		1552	4.23
Msp	*Myriophyllum spicatum*				0.61	1.35	11.83	4.17	20.64		59.11	2.86	216	8.85
Myrver	*Myriophyllum verticillatum*								1.07				3	------
Nf	*Najas flexilis*				0.53	1.08	5.91						22	------
Nitella	*Nitella* spp.		7.75	1.56	21.53	57.57	38.17		16.37	4.95	14.78		686	5.38
Nul	*Nuphar lutea*	0.45	4.23	0.78	2.73	14.05	1.61	89.58	16.01		6.40	2.86	214	6.92

Code	Taxon	A	B	C1	C2	D	E	F	G	H	I	J	Occ.	PLEX
Nus	*Nuphar lutea* x *pumila*				0.38	1.35	0.54		0.71		0.49		14	------
Nup	*Nuphar pumila*	0.45	3.99	0.39	1.29	4.59		2.08	0.71		0.99		58	5.38
Na	*Nymphaea alba*	6.76	45.07	1.17	35.78	24.05	11.83	50.00	16.01	0.99	4.43		872	3.08
Nyp	*Nymphoides peltata*		0.23		0.15		0.54		1.07		1.97		11	------
Oa	*Oenanthe aquatica*	0.45						4.17	3.20				12	------
Pama	*Persicaria amphibia*		1.41		0.68	9.19	20.43	41.67	46.98	0.99	32.51		306	7.95
Pil	*Pilularia globulifera*		0.23		0.68	2.70	2.15	2.08					25	5.38
Pal	*Potamogeton alpinus*		2.58	2.34	3.94	16.22	14.52		3.56	0.99	0.99		169	5.38
Pxgr	*Potamogeton alpinus* x *praelongus*					0.27							1	------
Pbe	*Potamogeton berchtoldii*		2.58	0.78	5.53	33.51	30.65	10.42	30.60	2.97	27.09		416	7.69
Pcol	*Potamogeton coloratus*				0.08	0.27	0.54		0.71		1.97		9	------
Pcom	*Potamogeton compressus*					0.27							1	------
Pcr	*Potamogeton crispus*		0.23		0.08	7.84	8.60	4.17	26.33	1.98	33.99		194	7.95
Pep	*Potamogeton epihydrus*				0.08								1	------
Pfil	*Potamogeton filiformis*		0.23		0.53	0.54	41.40		0.36	1.98	37.93	5.71	169	7.69
Pfr	*Potamogeton friesii*					1.61			1.42		10.34		28	9.23
Pgr	*Potamogeton gramineus*	0.45	1.64		6.67	11.89	54.84		2.49		14.29		278	7.31
Pxz	*Potamogeton gramineus* x *lucens*				0.76	1.89	2.69		0.36		1.97		27	7.69
Pxsp	*Potamogeton gramineus* x *natans*				0.08	0.27	0.54				0.49		4	------
Pgr/pxn	*Potamogeton gramineus* x *nitens*				0.15	0.27							3	------
Pxn	*Potamogeton gramineus* x *perfoliatus*		0.47		1.97	8.11	26.88		0.71		8.37		127	7.69
Plu	*Potamogeton lucens*				0.30	1.08	3.23	2.08	2.49		5.91		34	7.88
Pn	*Potamogeton natans*	1.30	42.96	10.94	70.66	53.78	58.60	8.33	42.70	6.93	28.57	8.57	1647	4.23
Pxg	*Potamogeton natans* x *polygonifolius*										0.49		1	------
Pob	*Potamogeton obtusifolius*		3.76		0.91	16.49	5.91		25.98		3.94		181	6.54
Ppec	*Potamogeton pectinatus*				0.38	0.27	18.28	12.50	15.66	1.98	53.69	54.29	220	8.85
Pper	*Potamogeton perfoliatus*		0.70		10.54	21.35	68.82	2.08	9.61	1.98	34.98		450	7.69
Ppol	*Potamogeton polygonifolius*	7 66	72.54	20.70	77.79	35.41	37.63	2.08	3.56	3.96	4.43	8.57	1633	3.08
Ppra	*Potamogeton praelongus*		0.23		3.03	4.32	15.05				1.48		88	5.38
Ppu	*Potamogeton pusillus*		0.47		0.38	2.43	13.44	6.25	16.73	0.99	46.80		187	7.95
Pr	*Potamogeton rutilus*				0.08		3.76						8	------
Pt	*Potamogeton trichoides*								3.20		0.99		11	------
Ra	*Ranunculus aquatilis* sens. str.		0.47		0.08	3.24	3.23	2.08	6.41	9.90	7.88		66	7.95
Rb	*Ranunculus baudotii*		0.23		0.15	0.27	14.52		0.36	10.89	24.63	11.43	97	7.69
Rc	*Ranunculus circinatus*					1.08			7.47		11.33	2.86	47	8.85
Rh	*Ranunculus hederaceus*		0.47		0.08	4.59	3.76	4.17	8.90	5.94	4.93	11.43	74	7.69
Ro	*Ranunculus omiophyllus*		0.23		0.38	0.81		4.17	0.71	0.99	1.97		18	------
Rp	*Ranunculus peltatus*		0.23		0.23	7.30	0.54		9.25	7.92	4.43		75	7.69
Rpse	*Ranunculus penicillatus* subspecies *pseudofluitans*										0.49		1	------
Rtr	*Ranunculus trichophyllus*		0.23		0.08	2.43	6.45		3.20	1.98	10.34		55	7.69
Ricflu	*Riccia fluitans*		0.70					4.17	1.42				9	------
Ruc	*Ruppia cirrhosa*						1.08		0.36		0.49		4	------

Code	Taxon	A	B	C1	C2	D	E	F	G	H	I	J	Occ.	PLEX
Rum	*Ruppia maritima*				0.08	0.54					2.46	60.00	29	------
Spa	*Sparganium angustifolium*	16.67	16.67	65.63	48.45	43.24	32.26		1.07	0.99	0.49		1140	4.23
Spema	*Sparganium emersum*		5.87		2.12	11.35	5.38		11.39	0.99	8.37		155	7.50
Sphag	*Sphagnum* (aquatic indet.)	100.00	60.33	54.69	22.74	15.41	4.30	10.42	2.85	2.97	1.97		1004	1.54
Spn	*Sparganium natans*	1.35	10.80	1.95	10.46	8.92	3.23	6.25	1.07	0.99	0.49		239	3.08
Spo	*Spirodela polyrhiza*		0.23				2.08		0.71		0.49		5	------
Sub	*Subularia aquatica*		0.23	6.25	20.17	16.22	5.38						353	4.23
Um	*Utricularia minor*	7.21	22.30	2.73	16.60	2.43	7.53		0.71	0.99	0.49		364	3.08
Uti	*Utricularia intermedia* sens. lat.	0.45	8.69	0.78	15.09	2.43	3.23		0.71				256	3.08
Uva	*Utricularia vulgaris* sens. lat.		5.40	0.39	8.95	6.22	6.99	2.08	2.49		0.49		187	4.23
Woa	*Wolffia arrhiza*								0.36				1	------
Zan	*Zannichellia palustris*		0.70			0.27	6.99	31.25	14.23		34.98	5.71	145	8.85

Notes. Occ. = the number of occurrences of each species in the dataset

Annex C: Relationships between the environmental variables, taxon richness and PLEX.

	Taxon Richness	Altitude	Surface area	Conductivity	pH	Alkalinity	PLEX
Taxon Richness	1.000	-0.205	0.163	-0.083	0.337	-0.083	0.224
Altitude	-0.206	1.000	-0.049	-0.110	-0.314	-0.254	-0.385
Surface area	0.163	-0.049	1.000	-0.013	0.017	-0.034	0.036
Conductivity	-0.083	-0.110	-0.013	1.000	0.145	0.226	0.166
pH	0.337	-0.314	0.017	0.145	1.000	0.453	0.680
Alkalinity	-0.083	-0.254	-0.034	0.226	0.453	1.000	0.593
PLEX	0.224	-0.385	0.036	0.166	0.680	0.593	1.000

Table C.1. Pearson correlation coefficients between environmental variables, taxon richness and PLEX. Coefficients in bold are significant values at the 0.05 level of significance (two-tailed test). Calculations performed using pairwise deletion.

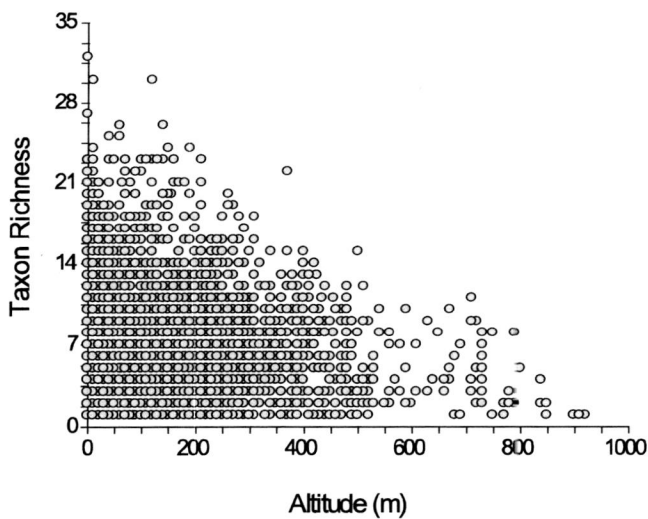

Figure C.1. Relationship between taxon richness and altitude.

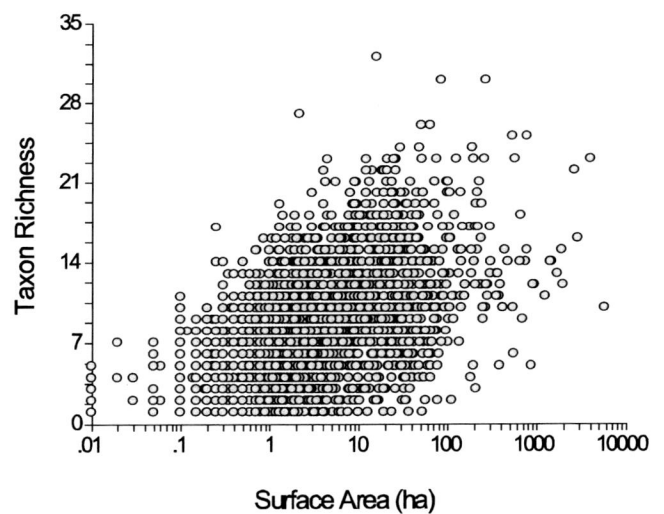

Figure C.2. Relationship between taxon richness and surface area.

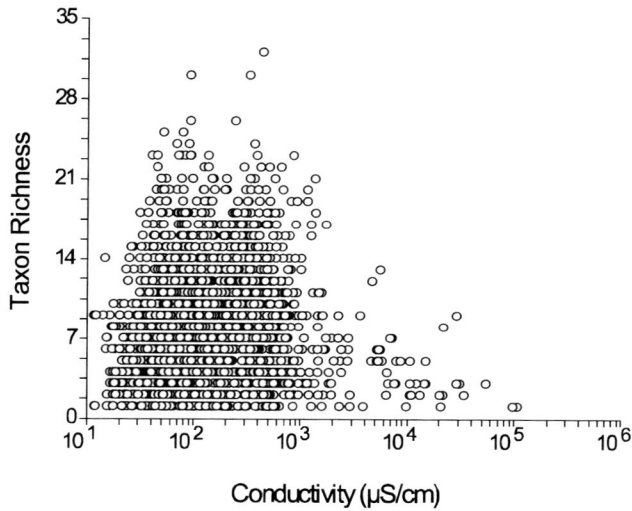

Figure C.3. Relationship between taxon richness and conductivity.

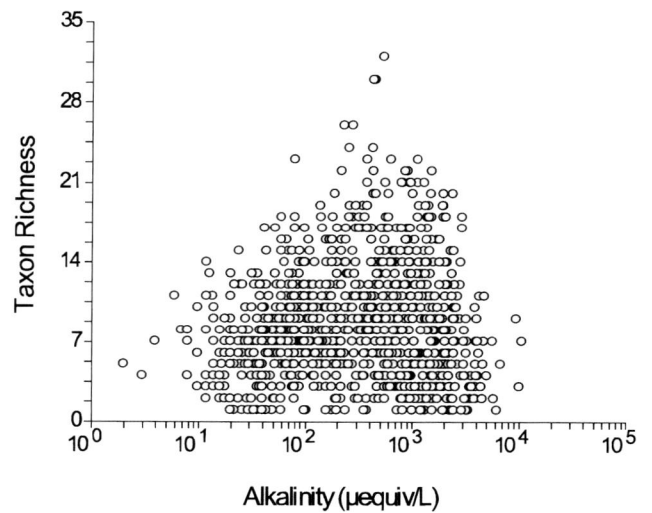

Figure C.5. Relationship between taxon richness and alkalinity.

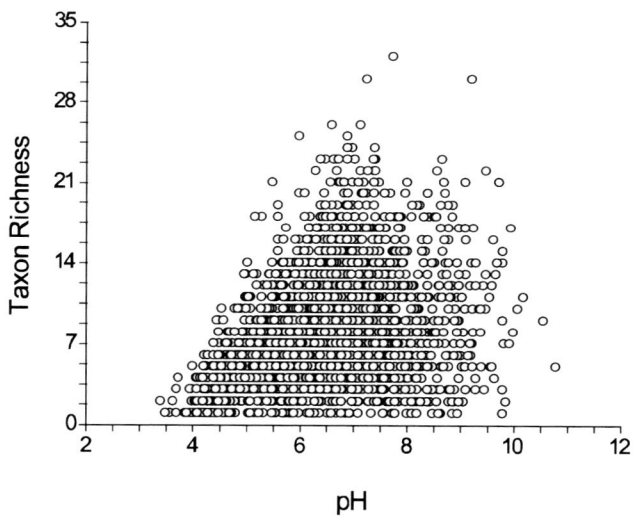

Figure C.4. Relationship between taxon richness and pH.

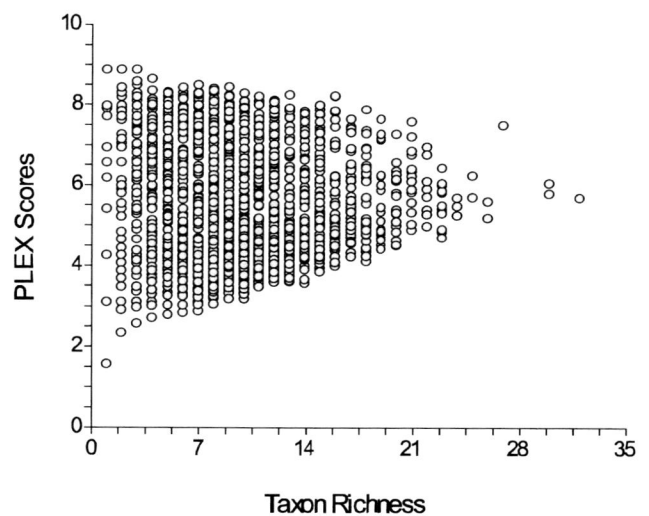

Figure C.6. Relationship between taxon richness and PLEX.

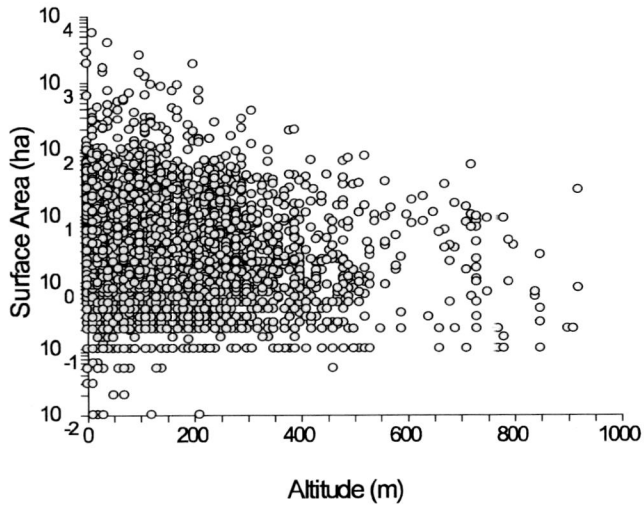

Figure C.7. Relationship between altitude and surface area.

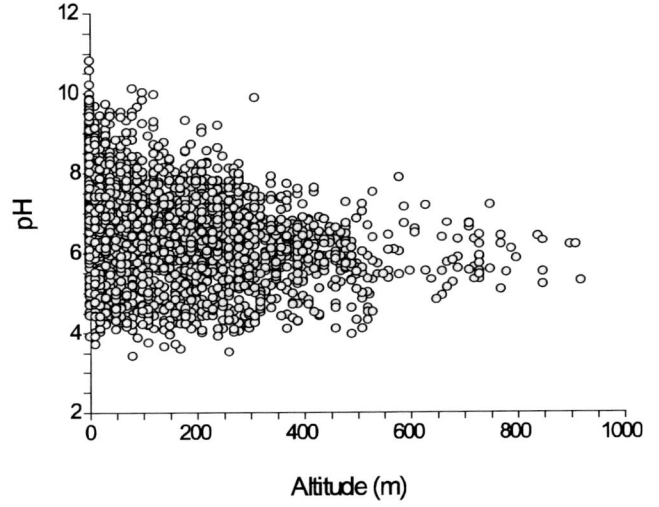

Figure C.9. Relationship between altitude and pH.

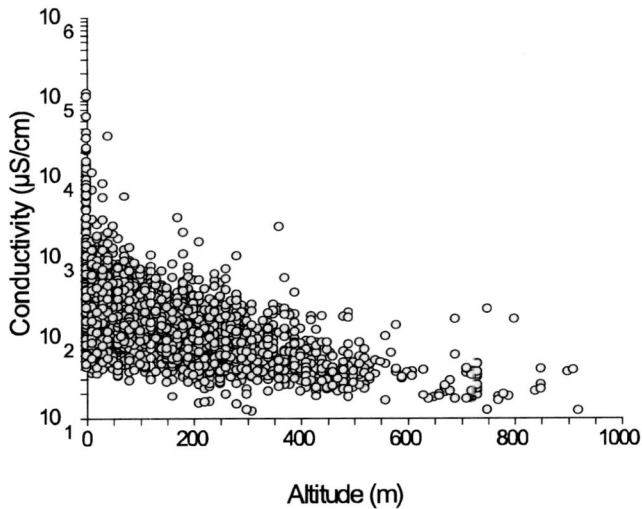

Figure C.8. Relationship between altitude and conductivity.

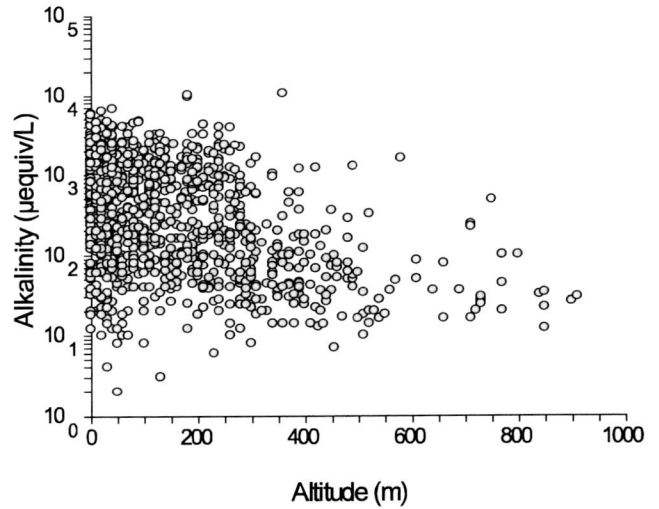

Figure C.10. Relationship between altitude and alkalinity.

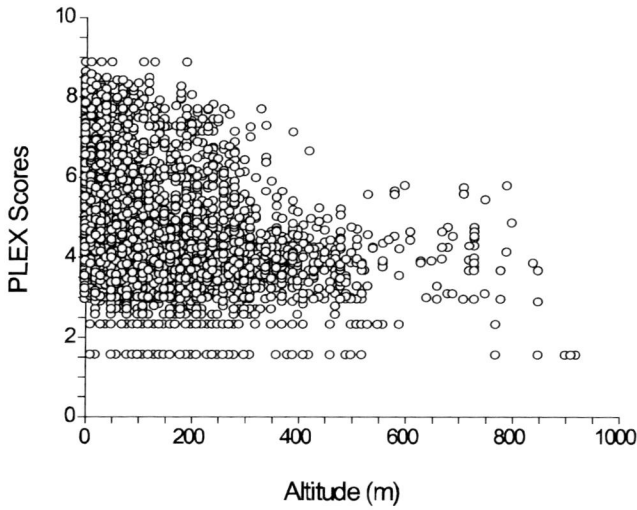

Figure C.11. Relationship between altitude and PLEX.

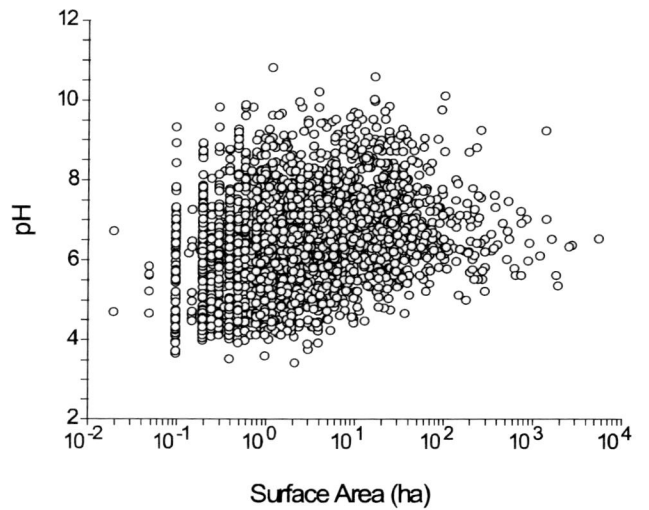

Figure C.13. Relationship between surface area and pH.

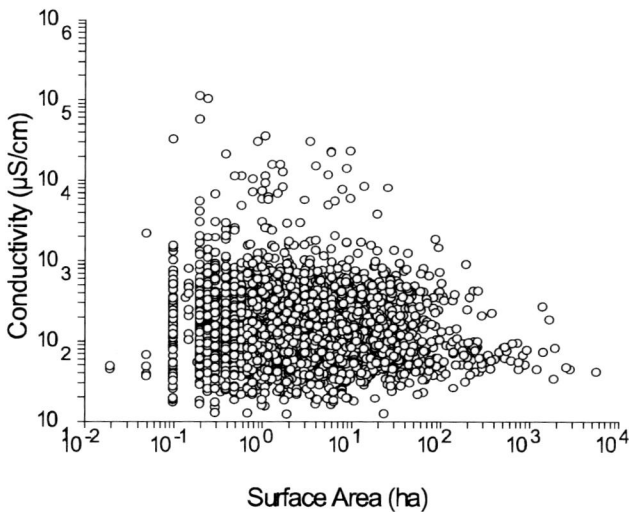

Figure C.12. Relationship between surface area and conductivity.

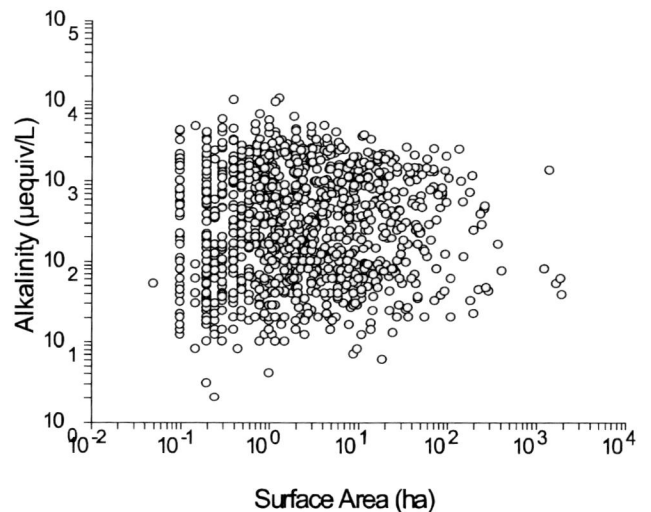

Figure C.14. Relationship between surface area and alkalinity.

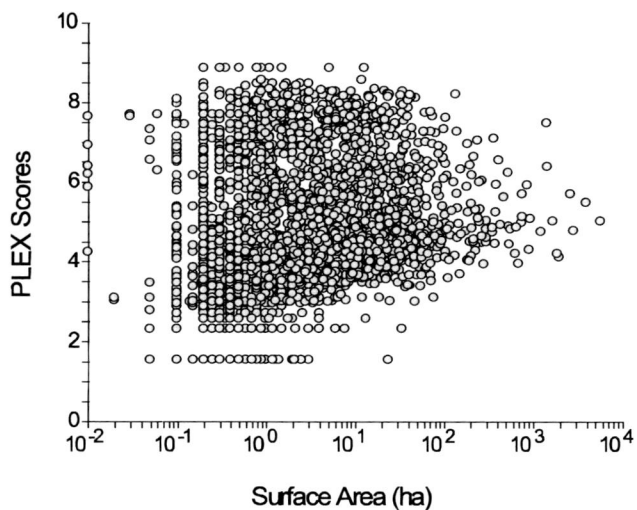

Figure C.15. Relationship between surface area and PLEX.

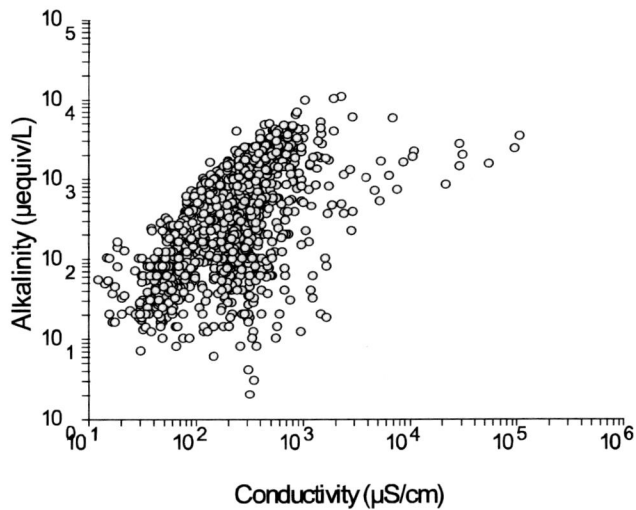

Figure C.17. Relationship between conductivity and alkalinity.

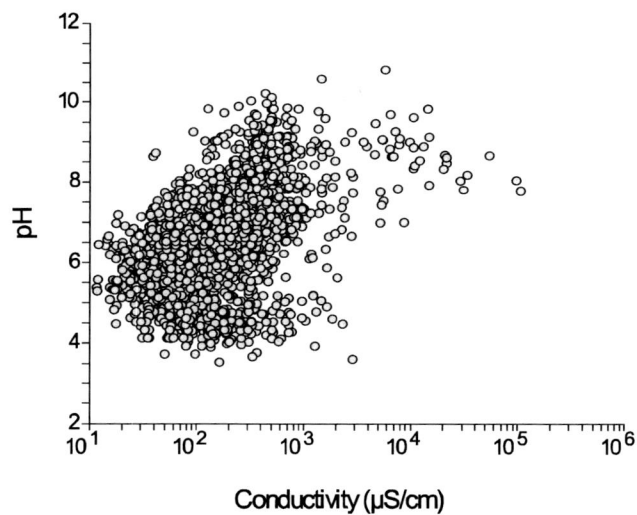

Figure C.16. Relationship between conductivity and pH.

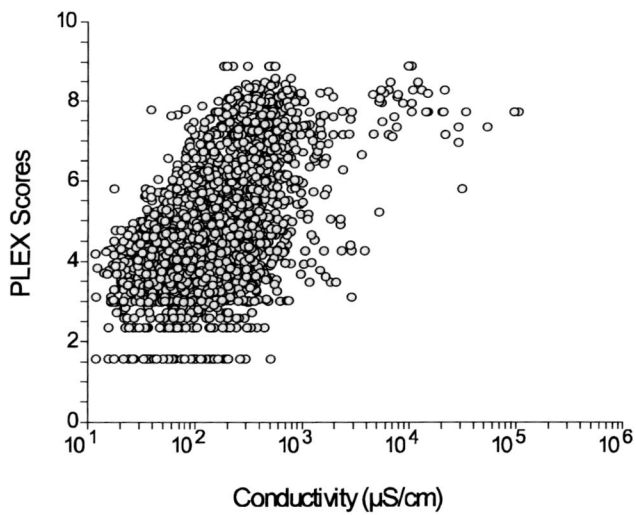

Figure C.18. Relationship between conductivity and PLEX.

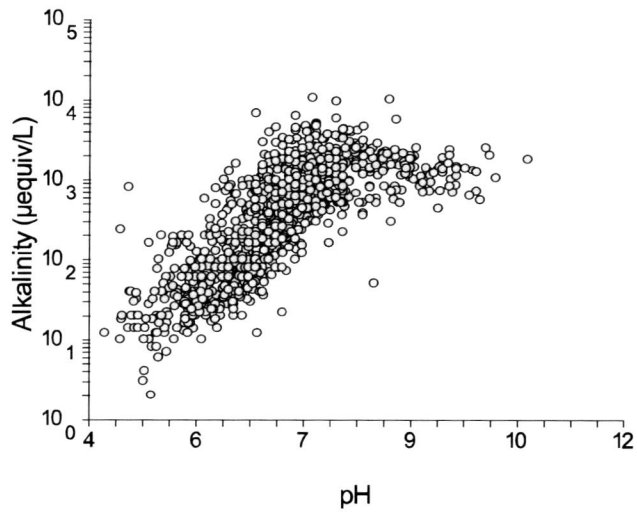

Figure C.19. Relationship between pH and alkalinity.

Annex D: Sites included in this study

Site code	Site name	Grid Reference	Lake Group	PLEX Score		Site code	Site name	Grid Reference	Lake Group	PLEX Score
Wales						CWP38W	Cotswold Water Park	SU0394	I	8.08
CW0001	Llyn Idwal	SH646595	C2	4.49		CWP40W	Cotswold Water Park	SU0393	I	8.05
CW0002	Llyn Cwellyn	SH560550	C2	4.07		CWP50W	Cotswold Water Park	SU0194	I	7.35
CW0003	Llyn Coron	SH368700	G	7.21		CWP54W	Cotswold Water Park	SU0094	G	8.40
CW0004	Llyn Dinam	SH311775	I	7.34		CWP9BE	Cotswold Water Park	SU1799	I	8.85
CW0005	Llyn Penrhyn	SH315770	G	7.79		EN0001	Abbots Moss	SJ590690	A	2.56
CW0006	Bugeilyn	SN822923	C2	4.95		EN0003	Airy Hill Pond	NY554314	D	6.19
CW0007	Llyn Eiddwen	SN605670	D	4.88		EN0004	Alcock Tarn	NY349079	D	5.26
CW0008	Llyn Fanod	SN603643	D	5.17		EN0005	Alkmond Park Pool	SJ480160	G	8.27
CW0009	Llyn Glanmerin	SN755991	D	4.64		EN0006	Allan Tarn	SD292894	D	6.10
CW0010	Llyn Gynon	SN800647	C2	4.94		EN0008	Angle Tarn:Bowfell	NY240070	C1	3.85
CW0011	Llyn Hir	SN789677	C2	4.72		EN0009	Angle Tarn:High Street	NY410140	C2	4.32
CW0012	Llynnoedd Ieuan	SN795815	C2	3.65		EN0010	Aqualate Mere	SJ773205	F	7.63
CW0013	Maes-Llyn	SN693628	D	6.26		EN0011	Baddiley East Mere	SJ595502	G	7.32
CW0014	Talley Upper	SN633332	D	6.18		EN0012	Bagmore Pit	TL860922	G	7.92
CW0015	Talley Lower	SN632337	G	6.01		EN0013	Bar Mere	SJ530470	F	8.08
CW0016	Kenfig Pool	SS790820	I	7.67		EN0014	Barfield Tarn	SD100860	C2	4.89
CW0017	Llyn Llech Owain	SN569151	B	3.65		EN0015	Barngates Tarn	NY300000	G	6.09
CW0018	Llyn Fach	SN905370	C2	4.09		EN0016	Barrow Plantation Tarn	SD410940	D	4.94
CW0019	Llanbwchllyn	SO118464	D	6.41		EN0017	Bassenthwaite Lake	NY215295	D	6.21
CW0020	Llangorse Lake	SO132265	I	7.25		EN0018	Baysbrown Tarn	NY310040	B	3.46
CW0021	Llyn y Fan Fawr	SN831217	D	4.62		EN0020	Beacon Tarn	SD274904	C2	4.64
CW0022	Hanmer Mere	SJ453392	F	7.52		EN0021	Berrington Pool	SJ520070	F	6.90
CW0023	Llyn Tegid	SH910335	D	6.53		EN0022	Berry Mound Sand Pit	SP097781	I	7.75
CW0024	Llyn Alwen	SH898567	C1	3.69		EN0023	Berth Pool	SJ430230	F	8.21
CW0025	Llyn Glasfryn	SH404422	D	6.31		EN0024	Betley Mere	SJ740470	G	7.81
CW0026	Llyn Rhos-ddu	SH425648	I	8.25		EN0025	Betton Pool	SJ511078	G	7.38
CW0027	Llynnau Mymbyr	SH642547	C2	4.23		EN0026	Biddulph's Pool	SK032097	B	5.44
CW0028	Gloyw Lyn	SH647298	C2	3.46		EN0027	Marbury Big Mere	SJ559454	G	7.16
CW0029	Llyn yr Wyth Eidion	SH470820	F	6.71		EN0028	Birchgrove Pool	SJ430230	F	7.82
CW0030	Llyn Cau	SH715124	C1	3.85		EN0030	Blackbrook Reservoir	SK455175	C2	5.63
CW0031	Llyn Llagi	SH649483	C1	3.93		EN0031	Blake Mere	SJ410330	F	7.85
CW0032	Llyn Ogwen	SH660605	C2	4.44		EN0032	Blea Tarn:Eskdale	NY166010	C2	4.23
CW0033	Broad Pool	SS510910	B	3.97		EN0033	Blea Tarn:Langdale	NY290040	C2	4.27
CW0034	Llyn Conwy	SH779461	C2	3.65		EN0034	Blea Water	NY449108	D	5.19
CW0035	Llyn Helyg	SJ115774	E	6.41		EN0035	Blelham Fish Pond	NY363007	H	5.38
CW0036	Llyn Bedydd	SJ471392	F	6.92		EN0036	Blelham Tarn	NY360005	D	5.58
CW0037a	Pant-yr-Ochain Pool 1 & 2	SJ349531	B	5.51		EN0037	Boltons Tarn	SD440930	H	5.88
CW0037b	Pant-yr-Ochain Pool 3	SJ347534	G	8.16		EN0038	Bomere Pool	SJ499081	F	5.90
						EN0039	Booths Mere	SJ760780	G	7.34
England						EN0040	Boretree Tarn	SD350870	D	5.16
CWP16E	Cotswold Water Park	SU1999	I	7.82		EN0041	Borwick Fold	SD445965	D	5.69
CWP18W	Cotswold Water Park	SU0596	G	8.27		EN0042	Borwick Fold Tarn	SD445965	D	5.79
CWP22W	Cotswold Water Park	SU0595	I	7.90		EN0043	Boultham Mere	SK950710	G	7.35
CWP24W	Cotswold Water Park	SU0595	I	8.21		EN0044	Bowscale Tarn	NY330310	C2	3.65
CWP29W	Cotswold Water Park	SU0394	I	8.25						

Site code	Site name	Grid Reference	Lake Group	PLEX Score
EN0046	Buckenham Marshes	TG350050	G	8.18
EN0047	Budworth Mere	SJ650760	F	7.85
EN0048	Budworth Pool	SJ590650	G	8.26
EN0049	Burdesley Abbey Fish Pond	SP047688	G	7.44
EN0050	Burnmoor Tarn	NY180040	C2	3.65
EN0051	Burton Gravel Pits: 1	SK942743	G	7.79
EN0052	Burton Gravel Pits: 2	SK943740	G	8.85
EN0053	Burton Gravel Pits: 3	SK945739	G	7.26
EN0054	Burton Gravel Pits: 4	SK944738	G	7.20
EN0055	Burton Gravel Pits: 5	SK946739	G	7.60
EN0056	Burton Gravel Pits: 6	SK948740	G	7.68
EN0057	Wast Water	NY163061	C2	4.49
EN0058	Burton Gravel Pits: 7	SK948739	G	8.07
EN0059	Burton Gravel Pits: 8	SK945741	G	7.69
EN0060	Burton Gravel Pits: 9	SK944743	G	8.01
EN0061	Calthorpe Broad	TG410258	F	6.28
EN0063	Castle Donington: Flax Pool	SK443291	G	7.02
EN0064	Castor Hanglands Main Pond	TF119016	G	7.44
EN0066	Chapel Mere	SJ540510	F	7.97
EN0067	Kyre Pool 2	SO624635	F	7.48
EN0068	Claife:Robinsons Tarn	SD368981	B	2.98
EN0069	Claife:Rough Hows Lily Pond	SD367984	B	2.98
EN0070	Claife: Scale Tarn	SD373975	B	3.23
EN0071	Claife: Wray Mires Tarn	SD360970	B	4.93
EN0072	Cleabarrow	SD420960	G	6.71
EN0073	Clumber Park Lake	SK620730	G	8.27
EN0074	Cofton Reservoir	SP000750	G	7.12
EN0075	Cole Mere	SJ435335	I	7.77
EN0076	Cop Mere	SJ800290	G	7.69
EN0077	Crag Lough	NY760680	I	6.72
EN0079	Crook Tarns	SD460952	E	6.02
EN0080	Cropston Reservoir	SK546108	G	6.96
EN0081	Crose Mere	SJ430300	F	7.56
EN0082	Culverthorpe	TF021399	G	7.55
EN0083	Culverthorpe 2	TF017399	G	7.71
EN0084	Cunswick Tarn	SD480930	F	5.05
EN0086	Deeping Gravel Pits: East pit	TF180082	I	8.16
EN0087	Deeping Gravel Pits: West pit	TF180082	F	8.09
EN0088	Deer Park Mere	SJ540508	F	7.80
EN0089	Devoke Water	SD157969	C2	4.50
EN0090	Dock Tarn	NY270140	C2	3.93
EN0091	Doddington Pool	SJ710460	F	8.28
EN0092	Dozmary Pool	SX190740	E	5.87
EN0093	Eel Tarn	NY189019	B	3.65
EN0094	Elter Water	NY334041	C2	5.11
EN0096	Englemere Pond	SU900680	B	4.58
EN0097	Ennerdale Water	NY100150	C2	5.01
EN0098	Eskdale Green Tarn	NY140000	C2	3.97
EN0100	Fenemere	SJ445225	F	6.57
EN0101	Flass Tarn	NY129035	C2	5.38
EN0102	Flass Tarn West	NY129035	C2	4.69
EN0103	Fleet Pond	SU820550	B	5.87
EN0105	Fowl Mere	TL880890	I	8.16
EN0106	Ghyll Head Reservoir	SD398923	C2	5.40
EN0108	Great Meadow Pond	SU960700	I	8.85
EN0109	Greendale Tarn	NY147075	C2	4.20
EN0110	Grimley Brickpits: southern section	SO835605	G	7.39
EN0111	Grimley Pool	SO826608	G	7.44
EN0113	Grizedale Tarn	SD394886	D	4.52
EN0114	Hagg Pond	SD366981	B	3.88
EN0116	Harrop Tarn	NY310130	B	4.08
EN0117	Hartsholme Lake	SK940690	F	6.62
EN0118	Hatch Mere	SJ550720	F	7.51
EN0119	Hawes Water	SD470760	F	6.01
EN0120	Hayeswater	NY431121	D	4.46
EN0121	Hell Kettles	NZ281109	J	8.85
EN0122	Helton Tarn	SD419849	G	7.33
EN0124	High Arnside Tarn	NY332012	D	5.88
EN0125	High Crag Tarn	NY350010	D	5.24
EN0126	Hodsons Tarn	SD369983	C2	4.36
EN0127	Home Mere	TL893897	I	7.71
EN0128	Hornsea Mere	TA190470	I	8.20
EN0130	Houghton Regis Chalk Pit	TL010230	I	7.09
EN0131	Hurcott Pool	SO850770	F	7.88
EN0132	Kettle Mere	SJ410340	F	7.52
EN0133	King's Mere	SU812640	B	2.69
EN0135	Kyre Pool 1	SO625636	G	8.04
EN0136	Lambhowe Tarn	SD417919	D	5.98
EN0137	Langtoft Gravel Pits	TF110110	I	7.47
EN0138	Laughton Forest: East Pond	SK860995	G	7.82
EN0139	Laughton Forest: North Pond	SK860995	G	5.73
EN0140	Laughton Forest: South Pond	SK860995	G	5.73
EN0141	Laughton Forest: West Pond	SK860995	H	6.41
EN0143	Lingmoor Tarn	NY300050	C2	3.88
EN0144	Little Langdale Tarn	NY300030	C2	4.10
EN0145	Little Sea	SZ030844	A	4.42

Site code	Site name	Grid Reference	Lake Group	PLEX Score	Site code	Site name	Grid Reference	Lake Group	PLEX Score
EN0147	Longlands Pond	NY200300	I	7.74	EN0198	Pippingford Park: Lake 4	TQ449306	B	4.81
EN0148	Loughrigg Tarn	NY340040	D	5.86	EN0199	Pippingford Park: Lake 5	TQ446312	D	4.04
EN0149	Lower Bittell Reservoir	SP018741	G	7.42	EN0200	Pirton Pool	SO870470	G	7.82
EN0150	Maer Pool	SJ790380	G	7.92	EN0201	Podmore Pool	SO854779	G	7.80
EN0152	Marbury Little Mere	SJ562457	F	6.90	EN0202	Priest Pot	SD357978	G	6.35
EN0153	Marton Pool:Baschurch	SJ450230	G	7.36	EN0204	Queen's Mere	SU810650	A	4.62
EN0154	Marton Pool:Chirbury	SJ290020	I	7.56	EN0205	Quoisley Little Mere	SJ549455	F	7.77
EN0155	Melchett Mere	SJ755802	I	7.85	EN0206	Rainworth Lake	SK580580	G	5.85
EN0157	Mere Tarn	SD260710	G	7.77	EN0207	Rapley Lakes	SU890640	B	2.95
EN0158	Misson Line Bank	SK710960	E	6.86	EN0208	Rather Heath	SD484958	D	5.30
EN0161	Moss Eccles Tarn	SD372968	C2	4.51	EN0209	Red Tarn:Helvellyn	NY340150	D	5.58
EN0162	Nassington	TL070970	G	7.67	EN0210	Red Tarn:Wrynose	NY267037	C2	3.65
EN0163	Nassington	TL070970	G	7.96	EN0212	Rhydd Brickpits	SO830450	G	7.51
EN0164	Nene Washes: Whittlesey: Site 1	TL276990	G	7.13	EN0214	Ripon Ponds	SE300700	G	7.24
EN0165	Nene Washes: Whittlesey: Site 10	TL267986	G	7.37	EN0215	Rostherne Mere	SJ740840	F	8.11
EN0166	Nene Washes: Whittlesey: Site 2	TL276990	G	8.21	EN0216	Sandhurst Ponds	SU860600	D	6.18
					EN0217	School Knott Tarn	SD420970	D	5.14
EN0167	Nene Washes: Whittlesey: Site 3	TL287994	G	7.47	EN0218	Sea Bank Clay Pits	TF530800	I	8.19
EN0168	Nene Washes: Whittlesey: Site 4	TL308999	G	7.97	EN0219	Semer Water	SD920872	G	7.14
EN0169	Nene Washes: Whittlesey: Site 5	TF383023	G	7.88	EN0221	Sheridans Lagoon	SU793324	A	2.31
EN0170	Nene Washes: Whittlesey: Site 6	TL274985	G	7.33	EN0222	Harnsey Tarn	NX994068	G	7.21
					EN0223	Shomere Pool	SJ504079	F	5.10
EN0171	Nene Washes: Whittlesey: Site 7	TL268990	G	7.50	EN0224	Siddick Pond	NY000300	G	7.61
EN0172	Nene Washes: Whittlesey: Site 8	TL267990	G	8.61	EN0225	Skeggles Water	NY479034	D	5.54
					EN0226	Skelsmergh Tarn	SD530960	F	6.63
EN0173	Nene Washes: Whittlesey: Site 9	TL265990	G	7.46	EN0227	Slapton Ley	SX825439	F	6.50
EN0174	Newstead Abbey Park Lake	SK520540	G	6.41	EN0228	Sleaford	TF085450	I	8.12
					EN0229	Sling Pool	SO947778	G	7.24
EN0176	No Man's Bank	SK028095	B	4.79	EN0230	Smokershole	TL880916	G	6.86
EN0177	Norbury Meres	SJ558493	G	8.16	EN0231	Southlake Moor	ST370300	G	7.43
EN0179	Oak Mere	SJ570670	B	5.71	EN0232	Sprinkling Tarn	NY220090	C2	4.19
EN0180	Old Alresford Pond	SU590330	F	7.81	EN0234	Stibbington Gravel Pit 1	TL080990	I	8.85
EN0181	Oss Mere	SJ560430	F	6.18	EN0235	Stibbington Gravel Pit 2	TL080990	I	8.46
EN0182	Over Water	NY252351	D	6.16	EN0236	Stibbington Gravel Pit 3	TL080990	G	8.85
EN0183	Parkhurst Hill	SU820500	C2	5.08	EN0237	Stibbington Gravel Pit 4	TL080990	G	7.51
EN0191	Peckforton Mere	SJ540570	H	4.62	EN0238	Stibbington Gravel Pit 5	TL080990	I	7.73
EN0192	Petty Pool	SJ610700	F	7.59	EN0239	Stibbington Gravel Pit 6	TL080990	G	8.33
EN0194	Pick Mere	SJ680770	F	7.69	EN0240	Stibbington Gravel Pit 7	TL080990	G	7.40
EN0195	Pippingford Park: Lake 1	TQ450299	H	7.98	EN0241	Stibbington Gravel Pit 8	TL080990	J	7.88
EN0196	Pippingford Park: Lake 2	TQ449301	G	5.65	EN0242	Stickle Tarn	NY280070	C2	4.18
EN0197	Pippingford Park: Lake 3	TQ448303	H	5.27	EN0243	Stocking Pool	SO689708	G	6.69
					EN0244	Stover Lake	SX830750	B	4.94
					EN0246	Strumpshaw Marsh	TG340060	G	7.47
					EN0247	Styhead Tarn	NY220090	C2	4.04
					EN0248	Swanholme Lakes SSSI: pit 1	SK940684	E	6.85
					EN0249	Swanholme Lakes SSSI: pit 10	SK944686	I	7.10

Site code	Site name	Grid Reference	Lake Group	PLEX Score
EN0250	Swanholme Lakes SSSI: pit 11	SK943685	C2	4.85
EN0251	Swanholme Lakes SSSI: pit 12	SK939684	D	5.48
EN0252	Swanholme Lakes SSSI: pit 2	SK942683	I	7.42
EN0253	Swanholme Lakes SSSI: pit 3	SK941687	E	5.97
EN0254	Swanholme Lakes SSSI: pit 4	SK944688	E	6.62
EN0255	Swanholme Lakes SSSI: pit 5	SK946685	D	5.86
EN0256	Swanholme Lakes SSSI: pit 6	SK945686	C1	3.33
EN0257	Swanholme Lakes SSSI: pit 7	SK945687	A	2.95
EN0258	Swanholme Lakes SSSI: pit 8	SK945687	A	2.31
EN0259	Swanholme Lakes SSSI: pit 9	SK943686	G	7.74
EN0260	Swanholme Lakes: Large Railway Pit	SK948685	E	6.69
EN0261	Swanholme Lakes: Medium Railway Pit	SK946682	G	7.51
EN0262	Sweat Mere	SJ430300	F	7.44
EN0263	Tabley Mere	SJ720770	G	7.71
EN0265	Talkin Tarn	NY545587	E	6.60
EN0266	Tarn Dub	NY853287	C2	4.78
EN0267	Tarn Hows	NY330000	C2	4.75
EN0268	Tarn Hows: High House Tarn	SD339994	C2	4.23
EN0269	Tattershall Gravel Pits: Mr Webb's Pit	TF199595	G	7.84
EN0270	Tattershall Gravel Pits: Mr Windley's Pit 1	TF193597	G	7.58
EN0271	Tattershall Gravel Pits: Mr Windley's Pit 5	TF194596	G	7.08
EN0272	Tattershall Gravel Pits: Mr Windley's Pit 2	TF194599	G	8.03
EN0273	Tattershall Gravel Pits: Mr Windley's Pit 3	TF196595	G	7.68
EN0274	Tattershall Gravel Pits: Mr Windley's Pit 4	TF199594	G	6.31
EN0275	Tatton Mere	SJ750800	I	7.73
EN0276	Tetney Blow Wells: No 1	TA320000	G	8.19
EN0277	Tetney Blow Wells: No 2	TA320000	H	6.78
EN0278	Tetney Blow Wells: No 3	TA320000	G	7.53
EN0279	Tetney Blow Wells: No 4	TA320000	G	7.47

Site code	Site name	Grid Reference	Lake Group	PLEX Score
EN0280	The Mere: Ellesmere	SJ400350	G	8.13
EN0281	The Mere: Mere	SJ733819	G	7.32
EN0282	Thompson Water	TL915950	I	8.15
EN0283	Three Dubs Tarn	SD378974	C2	4.40
EN0284	Tindale Tarn	NY600580	I	7.15
EN0285	Uckinghall	SO860370	G	5.84
EN0286	Upper Bittell Reservoir	SP020753	I	7.43
EN0288	Upton Warren: Brickiln Pool	SO930670	G	7.75
EN0289	Upton Warren: Moors Pool	SO930670	G	7.97
EN0290	Upton Warren: Sailing Pool	SO930670	G	7.18
EN0291	Urswick Tarn	SD271744	G	7.62
EN0292	Warren Heath Ponds	SU770580	B	4.04
EN0293	Warren Marsh	TL872952	G	6.09
EN0294	Wasing Wood Ponds	SU570630	B	5.03
EN0297	West Mere	TL885965	I	7.47
EN0300	West Tofts Mere	TL844919	G	6.31
EN0303	Westwood Great Pool	SO880633	G	7.35
EN0305	Whins Pond	NY555308	G	6.45
EN0306	Whisby Gravel Pits: 1	SK909665	I	7.68
EN0307	Whisby Gravel Pits: 2	SK905664	I	7.40
EN0308	Whisby Gravel Pits: 3	SK915666	I	7.90
EN0309	Whisby Gravel Pits: 4	SK905667	I	8.08
EN0310	Whisby Gravel Pits: 5	SK908667	I	7.93
EN0311	Whisby Gravel Pits: Teal's Pit	SK900600	I	7.64
EN0312	White Mere	SJ415325	G	8.21
EN0313	Wilden Washing Pool	SO823729	G	8.06
EN0315	Wise Een Tarn	SD370976	C2	3.57
EN0316	Witherslack Fish Pond	SD430860	G	6.89
EN0317	Witley Court	SO770650	F	6.96
EN0318	Woodhow Tarn	NY130040	D	5.51
EN0319	Woolmer Pond	SU788319	A	1.54
EN0320	Wytham Wood Meadow SSSI	SP400000	G	7.92
EN0321	Yardley Chase Ponds	SP850540	G	5.72
EN0322	Yardley Chase: S end	SP850540	I	6.60
EN0323	Buttermere	NY185155	C2	4.68
EN0324	Derwent Water	NY260210	C2	4.87
EN0325	Ullswater	NY420200	E	5.90
EN0326	Windermere	SD395985	D	6.36
EN0327	Sprinkling Crag Tarn	NY228095	C2	3.54
EN0328	Rydal Water	NY355062	D	5.67
EN0329	Grasmere	NY340065	C2	4.54
EN0330	Ennerdale Water	NY115145	C2	4.81
EN0331	Crummock Water	NY160185	C2	4.99
EN0332	Coniston Water	SD300940	C2	5.60
EN0333	Quoisley Big Mere	SJ546455	F	6.62

Site code	Site name	Grid Reference	Lake Group	PLEX Score
SMP102	Broomlee Lough	NY790698	E	6.70
SMP124	Greenlee Lough	NY773695	E	6.50
SMP194	Esthwaite Water	SD3696	D	6.06
SMP196	Sunbiggin Tarn	NY676077	G	6.89
SMP251	Tring Reservoirs - Wilstone	SP905130	I	8.22
SMP365	Woodwalton Fen - Gordon's Mere	TL230844	G	7.33
SMP388	Aldershot MOD - Lake I	SU8250	C2	3.85
SMP389	Aldershot MOD - Lake 2	SU8250	C2	4.23
SMP390	Aldershot MOD - Lake 3	SU8250	C2	5.19
SMP548	Stodmarsh - Fordwich Lake	TR15	G	7.90
SMP549	Stodmarsh - Stour Lake	TR16	G	7.97
SMP550	Stodmarsh - Trenley Park Lake	TR26	G	8.26
SMP564	Priddy Pool 1	ST546515	B	2.56
SMP707	Malham Tarn	SD895665	I	6.93
SMP744	Hayle Kimbro Pool, Lizard	SW694170	C2	4.74
SMP774	Ruan Pool, Lizard	SW71	B	5.38
SMP775	Croft Pascoe Pool, Lizard	SW731198	B	4.84

Scotland

Site code	Site name	Grid Reference	Lake Group	PLEX Score
HP4011	Fugla Water	HP491013	C1	3.65
HP5003	Kirk Loch	HP533048	I	7.00
HP5004		HP536047	E	5.19
HP5032	Loch of Watlee	HP591054	E	6.07
HP5034		HP573030	C2	4.62
HP5056		HP592036	C2	4.23
HP5071	Easter Loch	HP597013	I	7.14
HP5073	Loch of Snarravoe	HP568015	E	6.82
HP5104	North Water	HP584133	C2	4.52
HP5105		HP587131	A	2.31
HP6001		HP607088	C2	4.23
HP6010		HP611052	C1	2.98
HP6011		HP612053	C2	3.46
HP6101		HP605165	C1	4.38
HP6102		HP604164	B	2.98
HP6103		HP606165	C1	4.38
HP6104		HP605162	A	1.54
HP6105	Loch of Cliff	HP601115	E	6.14
HT9310	Rossie's Loch	HT973395	C1	3.97
HT9311	Mill Loch	HT967387	E	6.39
HT9314		HT972380	C1	4.49
HT9401	Lochs o' da Fleck	HT947401	B	3.92
HT9402	Lochs o' da Fleck	HT949401	H	4.62
HU1401		HU186498	E	5.91
HU1402		HU191499	E	6.00
HU1404		HU198484	D	4.23

Site code	Site name	Grid Reference	Lake Group	PLEX Score
HU1504		HU187555	C1	2.95
HU1507	Loch of Watsness	HU175506	E	6.46
HU1611		HU165610	C1	4.10
HU1612		HU166611	A	4.04
HU1614	Gorda Water	HU167608	D	4.69
HU1619		HU183612	D	7.02
HU2403		HU202478	E	6.25
HU2413		HU225467	C2	5.13
HU2414	Loch of Kirkigarth	HU238497	E	6.10
HU2433		HU272444	B	3.75
HU2502	Loch of Collaster	HU208572	C2	5.22
HU25126		HU242518	C2	4.23
HU25130	Loch of Flatpunds	HU246520	C2	4.83
HU25137	Mill Loch	HU282550	C2	4.40
HU25148		HU279539	B	3.27
HU25150		HU237502	E	5.80
HU25155	Grass Water	HU284534	C2	5.19
HU25160	Hulma Water	HU294527	C2	4.76
HU2589	Lunga Water	HU234526	C2	4.10
HU2607	Loch of Reva	HU299606	C2	4.42
HU2708	Loch of Houlland	HU215791	E	5.67
HU2718	West Loch	HU217778	D	5.92
HU2818	Dandy's Water	HU250836	C1	5.00
HU2840	Horse Lochs	HU258829	B	2.98
HU3001		HU386096	I	6.68
HU3002		HU388091	J	8.27
HU3103	Loch of Spiggie	HU372168	E	6.54
HU3106	Loch of Hillwell	HU376139	I	7.91
HU3209	Loch of Vatsetter	HU377233	D	5.60
HU3319		HU370332	E	6.13
HU3321		HU394359	B	3.56
HU3326	Mill Pond	HU385334	D	4.74
HU3415		HU385445	I	7.69
HU3509	Loch of Clousta	HU315581	C2	4.42
HU3522	Turdale Water	HU306529	E	5.96
HU3525	Loch of Houster	HU342551	E	5.88
HU3527		HU343550	E	5.88
HU3639	Gilsa Water	HU307631	C2	4.23
HU3647	Maa Loch	HU300600	E	5.72
HU3650		HU338616	C2	5.24
HU3740	Punds Water	HU326756	D	5.36
HU38103	Swabie Water	HU310854	C1	3.00
HU38108	Maadle Swankie	HU325869	C1	3.00
HU38119	Mill Lochs of Sandvoe	HU338882	C2	3.46
HU38124	Roer Water	HU337862	B	2.98
HU3813		HU327893	C1	3.52
HU38133	Mill Lochs of Sandvoe	HU341878	C2	3.46
HU38144		HU311838	C1	3.46

Site code	Site name	Grid Reference	Lake Group	PLEX Score
HU38240	Mill Lochs of Sandvoe	HU341882	C2	3.46
HU3837		HU334895	C2	3.94
HU3881	Muckle Lunga Water	HU327882	C1	3.00
HU3907	Innis Loch	HU332906	C1	4.52
HU4104		HU412174	D	6.54
HU4207		HU460236	D	5.53
HU4303	Lang Lochs	HU429380	C2	5.35
HU4423	Loch of Tingwall	HU417427	E	6.30
HU4443	Loch of Clickimin	HU464410	I	7.59
HU4512	Sand Water	HU415545	E	5.48
HU4528	Loch of Girlsta	HU433523	D	5.38
HU4532	Loch of Kirkabister	HU495583	B	3.69
HU4535	Loch of Benston	HU462535	C2	5.00
HU4611	Loch of Voe	HU415627	E	5.31
HU4712	Mussel Loch	HU472788	D	5.19
HU4801		HU443898	D	5.14
HU4804		HU441872	C1	3.23
HU4822		HU453845	B	3.00
HU4841		HU451825	C1	4.23
HU4842		HU452825	C1	4.23
HU4933	Hulk Waters	HU447946	C2	3.00
HU4934	Hulk Waters	HU447944	C2	3.46
HU4958	Loch of Birriesgirt	HU442910	C1	4.06
HU4961	Loch of Windhouse	HU496940	B	3.69
HU4964	Waters of Raga	HU475912	B	3.56
HU4965	Waters of Raga	HU476912	B	3.56
HU4966	Waters of Raga	HU477912	B	3.56
HU5301	Loch of Grimsetter	HU518398	E	6.15
HU5408	Ullins Water	HU521408	C2	3.75
HU5409		HU550408	D	5.00
HU5617	East Loch of Skaw	HU594665	D	4.23
HU5619	Loch Isbister	HU575643	D	5.38
HU5713	Loch of Grutwick	HU502707	C1	4.17
HU5803	Loch of Vasetter	HU533892	D	4.90
HU5905	Loch of Basta	HU507950	B	3.92
HU6601		HU603668	B	4.49
HU6802	Loch of Funzie	HU654898	E	6.63
HU6903	Skutes Water	HU622919	E	5.51
HU6908	Waters of Cruss	HU635910	E	5.15
HU6909	Waters of Cruss	HU636908	E	6.54
HU6918		HU666919	B	2.98
HU6920		HU667909	C2	4.69
HU6921		HU670908	C2	4.95
HY1001	Loch of Grutfea	HY193008	A	2.88
HY2001	Sandy Loch	HY219031	C2	5.49
HY2002	South dam*	HY232034	D	5.38
HY2003	Water of the Wicks	HY289002	A	1.54
HY2101	Loch of Clumly	HY255164	E	6.47
HY2102	Loch of Skaill	HY243182	I	7.73
HY2103	Mill Dam of Rango	HY262183	I	7.10
HY2104		HY256171	B	5.38
HY2105	Stromness Reservoir*	HY239107	E	6.88
HY2109		HY258161	E	5.96
HY2113		HY256172	B	5.38
HY2201	Loch of Banks	HY275234	E	5.99
HY2202	Loch of Boardhouse	HY268259	I	7.02
HY2203	Loch of Hundland	HY295261	E	6.44
HY2204	Loch of Sabiston	HY293223	I	7.33
HY2205	Loch of Isbister	HY257236	E	6.63
HY2206	The Loons	HY247242	I	6.89
HY2207	The Loons	HY253241	I	6.73
HY2208		HY257244	B	6.09
HY2211		HY257242	E	5.88
HY2212		HY259243	I	7.05
HY2216	The Loons	HY248241	I	6.89
HY2227		HY252227	I	6.62
HY2228		HY249242	E	5.96
HY3001	Loch of Kirbister	HY369079	I	7.31
HY3101	Loch of Bosquoy	HY307185	I	7.38
HY3102	The Shunan	HY307194	I	7.15
HY3103	Parro Shun*	HY305194	E	6.01
HY3104	Loch of Brockan	HY395192	I	7.29
HY3105	Loch of Wasdale	HY343148	E	6.78
HY3124	The Shunan	HY306196	I	7.15
HY3201	Peerie Water	HY399294	D	5.52
HY3202		HY393293	B	3.69
HY3203	Loch of Swannay	HY312283	E	6.62
HY3205	Loch of Vastray	HY399254	I	7.24
HY3206	Looma Shun	HY362237	A	3.72
HY3207	Lowrie's Water	HY344257	B	5.64
HY3208	Peerie Water	HY336272	E	6.91
HY3301	Loch of Moan*	HY382332	D	5.60
HY3302	Loch of Wasbister	HY396334	I	7.85
HY3303	Loch of Sacquoy	HY384351	I	7.69
HY3304	Muckle Water	HY395301	D	6.18
HY3401	Loch of the Stack	HY396494	J	8.08
HY4001	Loch of Ayre*	HY469012	J	8.27
HY4002	Loch of Graemeshall*	HY488020	I	7.64
HY4003	Loomi Shun	HY483060	H	7.69
HY4009	Loomi Shun	HY483058	B	3.08
HY4010	Black Loch	HY485056	H	7.69
HY4101	Little Vasa Water*	HY475187	H	7.69
HY4102	Vasa Loch	HY472184	I	8.46
HY4103	Peerie Sea	HY445111	J	8.27
HY4104	Resr.	HY424120	H	7.00
HY4105		HY476135	H	7.29

75

Site code	Site name	Grid Reference	Lake Group	PLEX Score
HY4201	Loch of the Graand	HY474272	I	7.44
HY4202	Manse Loch	HY477298	I	7.82
HY4203		HY445214	B	5.58
HY4204		HY433219	H	7.36
HY4205		HY435219	H	7.69
HY4206		HY451220	H	7.02
HY4207		HY447221	H	7.24
HY4211		HY451222	H	6.54
HY4214		HY473272	J	6.54
HY4301	Loch of Watten	HY478310	I	6.71
HY4302	Loch of Welland	HY477313	I	7.67
HY4303	Loch of Loomachun	HY400308	D	4.10
HY4304	Loch of Scockness	HY450329	I	7.72
HY4401	Loch of Burness	HY429481	I	7.75
HY4402	Loch of Garth*	HY470449	H	7.69
HY4403	Muckle Water	HY429431	I	7.93
HY4404	Loch Saintear	HY439475	I	8.06
HY4405	Loch of Swartmill	HY477458	I	7.88
HY4406		HY489426	H	7.79
HY4502	Craig Loch	HY446515	I	7.12
HY4503		HY493502	I	7.26
HY4504	Loch of St.Tredwell	HY494508	I	7.98
HY4505		HY497532	I	7.88
HY4506		HY495523	H	7.69
HY4507		HY496541	E	6.51
HY4508		HY499552	H	7.63
HY4509		HY498553	I	6.68
HY4510		HY499555	E	6.78
HY4512		HY446519	J	8.85
HY4514		HY447517	E	5.96
HY4515		HY495539	E	6.39
HY4516		HY496539	E	5.87
HY4517		HY496539	H	6.92
HY5001		HY546048	I	7.92
HY5003	Loch of Lakequoy	HY518084	I	7.69
HY5004	Loch of Messigate	HY518073	J	7.69
HY5005	Loch of Ouse	HY549072	I	8.15
HY5006	Loch of Tankerness	HY515095	E	6.99
HY5007		HY507095	I	7.91
HY5008		HY505084	E	6.25
HY5009		HY546070	H	6.54
HY5101	Lairo Water	HY509190	I	7.56
HY5102		HY505171	H	7.12
HY5103	Loch of Hestercruive	HY512111	I	7.69
HY5105		HY506171	H	7.12
HY5106		HY520173	H	7.82
HY5108		HY518167	H	6.83
HY5109		HY521167	D	5.77
HY5110		HY528165	H	7.60
HY5111	Loch of Sandside	HY520199	I	7.69
HY5301	Loch of Doomy	HY557342	I	7.69
HY5302	Loch of London	HY567348	J	7.69
HY5303	Mill Loch	HY564367	H	7.12
HY5304	Sealskerry Loch	HY533323	I	7.50
HY5501		HY500554	E	6.27
HY6201		HY683260	H	7.37
HY6202		HY650288	I	7.71
HY6203	Straenia Water	HY607227	H	6.54
HY6204		HY655279	H	7.12
HY6205	Meikle Water	HY665245	I	7.38
HY6206	Gricey Water	HY647261	I	7.93
HY6207	Blan Loch	HY635248	E	7.60
HY6208	Loch of Rothiesholm	HY624241	I	7.69
HY6209		HY622243	I	8.19
HY6210	Bruce's Loch	HY659282	H	7.71
HY6211		HY674248	I	7.97
HY6212	Lea Shun	HY663215	I	8.19
HY6213	Little Water	HY679236	I	7.17
HY6214	Loch of Matpow	HY642251	J	8.27
HY6215		HY613223	D	6.15
HY6216	Mill Loch	HY667296	I	7.76
HY6218		HY667241	I	7.64
HY6301		HY627389	H	7.12
HY6402	Roos Loch	HY657447	J	8.15
HY6403		HY664426	H	7.31
HY6406		HY668438	G	7.00
HY6407	Loch of Riv	HY685466	I	7.35
HY6408		HY686425	J	7.69
HY6409	Bea Loch	HY654400	I	7.93
HY6410		HY667400	J	8.46
HY6414		HY659418	J	8.85
HY7402	Loch of Langamay	HY746444	E	5.77
HY7403	North Loch	HY753458	I	8.05
HY7404	Loch of Rummie	HY756450	I	8.08
HY7405	Westayre Loch	HY726444	J	7.69
HY7501		HY753537	I	7.69
HY7502		HY753538	I	7.64
HY7503	Ancum Loch	HY763545	I	8.27
HY7504	Brides Loch	HY770522	I	8.27
HY7505		HY774522	H	7.50
HY7506		HY759526	H	7.69
HY7507	Dennis Loch	HY788554	I	7.50
HY7508	Loch of Garso	HY772554	I	7.98
HY7509	Loch Gretchen	HY749529	I	8.19
HY7510	Hooking Loch	HY764533	I	7.36
HY7511		HY786558	E	6.49

Site code	Site name	Grid Reference	Lake Group	PLEX Score
HY7512		HY782562	I	7.31
HY7514	Trolla Vatn	HY783562	J	8.85
HZ2701		HZ218738	C2	4.23
HZ2702		HZ221737	B	4.62
NA9116		NA982120	C2	5.13
NA9117		NA984120	C2	4.23
NA9201		NA986279	C1	2.98
NA9202		NA985278	C1	3.75
NA9203		NA984278	A	2.56
NA9204		NA984277	B	2.56
NA9210	Loch Greivat	NA987265	C2	4.35
NA9211		NA980263	C2	4.87
NA9218		NA985276	B	2.69
NB0048	Lochan Beag	NB055079	C2	4.04
NB0067	Loch a' Mhuilinn	NB089067	C2	3.77
NB0075	Loch an Duin	NB020014	C2	3.57
NB0080	Loch na Gaoithe	NB023000	C2	3.65
NB0086		NB000004	C2	3.57
NB01159	Loch na Cleavag	NB005136	C2	4.76
NB01165	Loch a' Ghlinne	NB025125	C1	4.23
NB01175	Loch Maolaig	NB063109	C2	4.18
NB01176	Loch Ashavat	NB072119	C2	4.18
NB0282	Loch na Clibhe	NB011247	C2	3.85
NB0283		NB006246	C2	3.50
NB03112		NB015312	C2	3.57
NB0314	Loch Mheacleit	NB049366	C2	4.09
NB0336		NB001323	C1	3.46
NB0337		NB002322	C1	4.23
NB0348		NB099357	B	3.82
NB0379		NB099346	C2	4.19
NB1028	Loch a' Mhorghain	NB153049	C2	3.92
NB11101		NB119132	C1	3.21
NB11102		NB120132	A	2.31
NB11103	Loch Chleistir	NB123132	C2	4.18
NB11144	Lochan Fheoir	NB103125	C2	3.91
NB11150		NB120132	B	3.08
NB1171	Loch Voshimid	NB104130	C2	4.32
NB1172		NB106131	C1	3.21
NB1205	Loch Sandavat	NB118290	C2	4.28
NB1304		NB103351	B	4.18
NB1340		NB153376	C2	3.87
NB1355	Loch a' Bhaile	NB196382	J	7.12
NB1370	Loch Baravat	NB155355	C2	4.04
NB1373	Lochan Sgeireach	NB151353	C2	4.44
NB1404	Loch an Duin	NB189408	D	5.30
NB2028		NB211041	C2	3.61
NB2030		NB214038	B	3.00
NB2039	Loch Beag	NB219028	C2	3.75

Site code	Site name	Grid Reference	Lake Group	PLEX Score
NB2141	Loch an Rathaid	NB224160	C2	3.90
NB2206		NB205286	C2	4.52
NB23169	Loch Smuaisaval	NB200300	C1	3.77
NB2333	Loch Laxavat Ard	NB246382	C2	4.38
NB2337	Loch Airigh Seibh	NB260388	C2	4.29
NB2352		NB261380	C1	3.94
NB2353		NB259381	C1	3.27
NB2391	Loch na Beinne Bige	NB223353	C2	4.49
NB2392	Loch Bharavat	NB223342	C2	4.38
NB2409	Loch Dalbeg	NB227457	D	5.03
NB2410	Loch Raoinavat	NB235460	C2	4.57
NB24109	Loch Ordais	NB283486	J	8.27
NB2415	Loch na Muilne	NB244467	D	4.65
NB2416	Loch a' Bhaile	NB255473	E	5.94
NB2429	Loch Tuamister	NB264456	B	3.08
NB3112		NB322197	C2	4.52
NB31335		NB377190	C2	4.51
NB3201	Loch Achmore	NB310285	E	6.11
NB3211	Loch Foid	NB312279	E	5.29
NB3212		NB314280	A	2.31
NB3213		NB312282	A	2.31
NB3214		NB313282	A	2.31
NB3253	Loch Orasay	NB390280	C2	4.38
NB33192	Loch na Linne	NB333302	C2	3.55
NB3374		NB336346	C1	4.81
NB3401	Loch Arnol	NB300490	D	5.38
NB3402		NB303487	E	5.15
NB3403		NB314497	C2	4.76
NB3404	Loch na Muilne	NB318497	C2	5.26
NB3407	Loch Mor Barvas	NB346500	E	6.58
NB3501	Loch Ereray	NB328505	D	6.18
NB3510	Loch an Duin	NB394543	E	5.38
NB4109	Loch a' Ghruagaich	NB412180	C2	3.72
NB4118		NB404168	C2	4.13
NB4268		NB407258	E	6.20
NB4324		NB459389	J	6.92
NB4325		NB458385	J	7.14
NB4331		NB460389	J	6.92
NB4345		NB458313	J	7.92
NB4347	Loch Branahuie	NB474324	H	7.12
NB44415	Loch Gunna	NB405414	C2	4.97
NB4503	Loch Baravat	NB462596	C2	4.47
NB4536	Loch Maravat	NB404536	C1	4.44
NB4606	Loch Dibadale	NB480613	G	7.00
NB5201	Loch Cuilc	NB506294	B	3.02
NB5202		NB509295	A	4.13
NB5301	Loch an Tiumpan	NB568370	E	6.18
NB5401	Loch Scarrasdale	NB503498	C2	4.20

Site code	Site name	Grid Reference	Lake Group	PLEX Score		Site code	Site name	Grid Reference	Lake Group	PLEX Score
NB5424		NB533498	C2	4.37		NC0252	Loch Dubh	NC073241	C2	4.55
NB55198		NB501504	A	2.56		NC0262		NC077238	C2	4.18
NB55199		NB500505	A	2.31		NC0264	Loch Culag	NC098217	C2	4.00
NB55200		NB500504	A	2.31		NC0266		NC077248	D	5.36
NB55201		NB500503	A	1.54		NC0267		NC090249	C2	3.57
NB5601		NB526644	G	6.77		NC0268		NC059257	B	3.69
NB5602	Loch Stiapavat	NB529642	D	6.10		NC0269		NC079270	C2	3.41
NB9105		NB977170	C2	3.57		NC0284		NC061260	C2	4.00
NB9106		NB978170	C2	4.62		NC0301	Loch Cul Fraioch	NC025330	C2	4.07
NB9107		NB982170	C2	3.65		NC0302	Loch an Achaidh	NC028337	B	4.06
NB9108	Loch Airigh Blair	NB985169	C2	3.85		NC0304		NC028335	B	3.58
NB9110	Loch na Totaig	NB980160	C2	4.81		NC0309	Loch Eileanach	NC090314	C2	3.92
NB9111		NB985164	C2	3.54		NC0310	Loch na Claise	NC035307	E	5.11
NC0005	Loch Bad na h-Achlaise	NC084088	C2	4.69		NC1003	Loch nan Ealachan	NC176090	C2	3.88
NC0101	Loch Rubha na Breige	NC073190	D	4.74		NC1004		NC181090	A	2.31
NC0102	Loch an Arbhair	NC079188	C2	4.52		NC1005	Lochan an Ais	NC185090	C2	4.23
NC0103	Loch a' Choin	NC083185	C2	4.64		NC1009	Lochan Fada	NC182085	C2	4.23
NC0105	Lochan Fada	NC086179	C2	3.59		NC1010	Clar Loch Beag	NC181080	C2	3.46
NC0114	Lochan Sal	NC071151	C1	4.42		NC1013	Clar Loch Mor	NC177073	C2	3.80
NC0119	Loch Buine Moire	NC098155	C2	3.65		NC1014		NC181074	C2	3.85
NC0126		NC086142	C2	3.97		NC1016		NC180070	C2	4.23
NC0127		NC088147	C2	3.97		NC1022	Loch Dhonnachaidh	NC175068	C2	3.65
NC0128	Loch Call an Uidhean	NC092145	C2	4.00		NC1111	Loch na Gainimh	NC176186	C2	4.56
NC0129		NC090142	C2	3.51		NC11132		NC169191	C2	4.23
NC0131		NC023131	B	3.85		NC1158	Loch Veyatie	NC180136	E	5.26
NC0132		NC030130	C2	3.51		NC12145	Loch an Alltain Duibh	NC145207	C2	3.96
NC0133		NC039135	C2	4.33		NC12159		NC154205	C2	3.57
NC0137	Loch na Dail	NC084135	C2	3.88		NC12167		NC198220	B	3.37
NC0138		NC089132	C2	3.97		NC1218	Loch Crocach	NC103273	C2	4.38
NC0139	Loch Uidh Tarraigean	NC094138	C2	3.80		NC1219	Loch Preas nan Aighean	NC113276	C2	4.09
NC0140	Polly Lochs	NC097136	C2	3.51		NC1227		NC104266	C2	3.73
NC0144	Loch Raa	NC018120	C2	4.50		NC1230		NC111265	C2	3.85
NC0145		NC061128	C2	4.23		NC1231	Loch an Tuir	NC116265	C2	3.92
NC0154	Loch Vatachan	NC018110	C2	4.07		NC1234	Loch Beannach	NC137262	C2	4.15
NC0159	Loch Bad a' Ghaill	NC080100	C2	4.46		NC1240	Loch na h-Innse Fraoich	NC163263	C2	4.02
NC0162		NC080182	B	2.69		NC1243	Loch Assynt	NC220244	C2	4.97
NC02121		NC087273	C2	3.46		NC1244	Loch an Tuirc	NC113259	C2	4.08
NC02140		NC092261	B	3.59		NC1249	Loch Bad nan Aighean	NC133255	C2	4.44
NC0215	Loch an Aigeil	NC041281	E	5.67		NC1251	Loch Uidh na Geadaig	NC142257	C2	3.73
NC0233		NC086272	C2	3.65		NC1253	Loch Torr an Lochain	NC163258	C2	4.11
NC0234		NC088273	C2	3.37		NC1258	Loch na Garbhe Uidhe	NC160249	C2	3.94
NC0235	Loch nan Lion	NC096274	C2	3.75		NC1260	Loch a' Mhuilinn	NC199242	C2	3.37
NC0244		NC082269	C2	3.58		NC1263	Loch a' Ghlinnein	NC169233	C2	3.90
NC0246		NC061264	B	3.37		NC1268		NC134222	C2	3.41
NC0247	Loch an Ordain	NC066260	C2	4.00		NC1269	Loch an Leothaid	NC143226	C2	3.76
NC0248		NC061251	B	4.58		NC1270	Loch Feith an Leothaid	NC185225	C2	4.23
NC0250	Lochan Sàile	NC075246	J	7.31		NC1271		NC197223	C2	3.80
NC0251	Manse Loch	NC095248	C2	4.50		NC1272	Loch Druim Suardalain	NC113217	C2	4.82

78

Site code	Site name	Grid Reference	Lake Group	PLEX Score
NC1275	Loch Crom	NC146219	C2	3.51
NC1276	Loch Bad an t-Sluic	NC152217	C2	3.50
NC1290		NC102267	B	3.52
NC1292		NC108270	B	3.88
NC1301	Loch a' Mhuillinn	NC165392	C2	4.30
NC1302	Loch a' Chreagain Daraich	NC170395	C2	3.42
NC1307	Loch Duartbeg	NC166388	C2	4.42
NC1312	Loch Duartmore	NC187373	C2	4.13
NC1314	Loch an Ruighein	NC186367	C2	3.90
NC1315		NC185362	C2	3.65
NC1318	Lochan na Dubh Leitir	NC174350	C2	3.65
NC1319		NC169348	C2	4.73
NC1320	Lochan Torr an Lochain	NC180343	C2	3.54
NC1321	Loch a' Meallard	NC156336	C2	4.12
NC1322	Lochan nan Gad	NC185335	C2	3.57
NC1324	Loch Drumbeg	NC116326	C2	4.76
NC1325	Loch Ruighean an Aitinn	NC126328	C2	4.13
NC1334	Loch Poll	NC101303	C2	4.36
NC1341		NC155305	C2	3.46
NC1343		NC185337	B	3.08
NC1344		NC166347	C2	4.00
NC1345		NC189334	B	3.16
NC1346		NC197375	C2	4.29
NC1347		NC189336	C2	3.51
NC1402	Loch Dubh	NC166488	D	5.71
NC1405	Loch nam Brac	NC180480	C2	3.65
NC1409	Clar Loch	NC187474	C2	4.23
NC1410		NC154468	C2	3.57
NC1411		NC154471	C2	3.85
NC1414	Loch a' Phreasain Challtuinne	NC186467	C2	3.77
NC1415	Lochain Bealach an Eilein	NC156464	C2	4.23
NC1416	Lochain Bealach an Eilein	NC157461	C2	3.65
NC1417		NC166461	C2	3.65
NC1418	Loch Laicheard	NC179461	C2	4.40
NC1423	Loch a' Bhadaidh Daraich	NC167446	C2	4.86
NC1424	Loch Leathad nan Cruineachd	NC161439	C2	3.57
NC1425	Lochan Sgeireach	NC166436	C2	3.51
NC1426	Lochan a' Mhuinean	NC156475	C2	4.44
NC1430	Loch a' Mhill Dheirg	NC152433	C2	4.42
NC1431	Loch an Daimh Mor	NC159431	C2	4.67
NC1432		NC162429	C2	4.13
NC1433	Loch a' Chreagain Thet	NC168423	C2	4.42
NC1434		NC184424	C2	3.77
NC1435		NC195426	C2	3.54
NC1436		NC187421	C2	3.37
NC1438		NC172414	C2	4.33
NC1441		NC161404	C2	4.66
NC1446		NC183462	C2	4.11
NC1447		NC170434	C2	3.31
NC1448		NC169439	C2	3.54
NC1449		NC155448	E	5.00
NC1450		NC169418	C2	3.57
NC1473		NC167457	B	3.08
NC1474		NC166456	C2	3.54
NC1502	Loch Poll a' Bhacain	NC188535	C2	3.46
NC1504		NC188532	C2	3.51
NC1513		NC194533	B	3.72
NC1601	Lochain nan Sac	NC198624	C2	3.74
NC1602	Loch a' Chreadha	NC186612	C1	3.65
NC1603		NC192603	C2	4.35
NC1604	Loch Aisir	NC198603	C2	4.96
NC1605		NC192633	C1	3.46
NC1606		NC185609	C2	3.94
NC1607		NC187611	C1	3.00
NC1608		NC187617	C2	4.41
NC1609		NC195622	C2	3.37
NC2044	Loch a' Chroisg	NC220025	C1	3.78
NC2053	Lochanan na Ceireag	NC213018	C2	4.09
NC2060		NC215013	C2	4.23
NC2061		NC212011	C2	3.85
NC2064		NC212010	C2	3.55
NC2065		NC238011	C2	3.19
NC2073	Lochanan nan Sailean Beaga	NC211008	C2	4.07
NC2076		NC212006	C2	3.97
NC2077		NC210005	C2	3.80
NC2081	Clar Lochan	NC242009	C2	3.77
NC2082	Clar Lochan	NC244007	C2	3.77
NC2083	Clar Lochan	NC246006	C2	3.31
NC2103	Loch Mhaolach-coire	NC276194	E	5.26
NC2105	Lochan Fada	NC206166	C2	3.88
NC2106	Loch a' Chroisg	NC219155	C2	4.18
NC2110	Loch na Gruagaich	NC242158	C2	3.85
NC2111	Loch Awe	NC246153	C2	5.07
NC2112		NC269153	E	5.47
NC2114	Feur Loch	NC270135	C2	4.48
NC2115	Loch Urigill	NC244100	E	5.22
NC2116	Loch Borralan	NC263107	D	5.02
NC2119		NC283178	C1	3.75
NC2123	Cam Loch	NC215136	E	5.30
NC2126		NC211161	C1	3.46
NC2144		NC278180	B	3.08
NC2201		NC227294	C2	3.65
NC2202	Loch nan Eun	NC231298	C2	3.65

Site code	Site name	Grid Reference	Lake Group	PLEX Score
NC2203	Loch na Gainmhich	NC244288	C2	3.65
NC2204	Loch Coire a' Bhaic	NC246294	C2	3.65
NC2206	Lochan Bealach Cornaidh	NC208281	C1	4.23
NC22107		NC288230	C1	4.23
NC2211	Loch nan Caorach	NC295275	C1	4.23
NC2213		NC218259	C1	3.46
NC2214	Lochan an Duibhe	NC219255	C2	3.65
NC2215	Lochan Feoir	NC229252	C2	3.46
NC2216	Loch Bealach na h-Uidhe	NC264256	D	5.54
NC2218		NC264240	D	4.62
NC2219		NC270241	D	4.72
NC2220	Loch Fleodach Coire	NC274248	D	5.16
NC2221	Loch nan Cuaran	NC291239	D	5.77
NC2222	Loch nan Caorach	NC290234	D	5.64
NC2223	Loch Meall nan Caorach	NC290230	B	3.08
NC2224		NC203220	C2	4.00
NC2274		NC203222	C2	3.85
NC2289		NC293277	C1	3.65
NC2297		NC275242	C1	3.65
NC23169		NC229303	C2	3.65
NC2318	Loch Allt na h-Airbe or Loch Yucal	NC205370	C2	3.57
NC2334		NC267360	C2	4.13
NC2336	Loch na Creige Duibhe	NC288356	D	5.28
NC2346		NC260350	C2	4.37
NC2349	Loch Unapool	NC227319	C2	4.23
NC2355	Loch Airigh na Beinne	NC219311	C2	3.70
NC2356		NC225312	C2	3.57
NC2359		NC202364	C2	3.65
NC2401		NC217495	C2	3.74
NC2402	Loch na Claise Luachraich	NC223496	C2	3.51
NC2403		NC229497	C2	4.00
NC2404		NC234495	C2	4.58
NC2406		NC276492	C2	3.72
NC2407	Blarloch Mor	NC286495	C2	3.56
NC2408	Loch Cul Uidh an Tuim	NC291493	C2	3.72
NC2409	Loch na Fiacail	NC232488	C2	3.59
NC2413	Loch a' Garbh-bhaid Mor	NC275482	C2	3.46
NC2414		NC287486	C2	3.51
NC2415		NC291484	C2	3.63
NC2416		NC289481	C2	3.59
NC2417		NC294481	C2	3.65
NC2419	Lochan an Fheidh	NC200477	C2	3.60
NC2426	Loch na Claise Fearna	NC200468	C2	4.41
NC24295		NC236420	C2	3.85
NC24307		NC246424	C2	3.65
NC2431	Loch Bad an t-Seabhaig	NC232457	C2	3.46
NC2439	Loch Airigh a' Bhaird	NC236453	C2	4.23
NC2447	Loch a' Cham Alltain	NC283446	C2	3.74
NC2448	Caol Lochan	NC284440	C2	3.57
NC2449	Loch an Nighe Leathaid	NC292445	C2	4.35
NC2451		NC209429	C2	3.94
NC2454	Lochain Doimhain	NC225430	C2	3.59
NC2457	Loch Stack	NC295420	C2	4.39
NC2458	Loch Grosvenor	NC281433	C2	3.57
NC2459	Clar Loch Mor	NC210426	C2	3.77
NC2461	Loch Eileanach	NC243425	C2	3.94
NC2462		NC299417	C2	3.35
NC2464	Loch na h-Ath	NC238418	B	3.08
NC2469		NC206431	C2	3.46
NC2493		NC204436	C2	3.65
NC2501	Loch Aisir Mor	NC215592	C2	4.73
NC2504	Loch na Larach	NC217582	C2	4.00
NC2505	Loch Carn Mharasaid	NC237584	C2	3.79
NC2506	Loch Innis na Ba Buidhe	NC226567	C2	4.23
NC2508	Loch Tarbhaidh	NC297557	C2	4.51
NC2509	Lochain Dubha	NC298551	C2	4.23
NC25100		NC232533	C2	3.92
NC2511	General's Loch	NC264540	C2	4.02
NC2513	Loch Eileanach	NC208534	C2	3.61
NC25133		NC248514	C2	3.65
NC25139		NC275536	C2	3.08
NC2514	Loch a' Phreasan Chailltean	NC213534	C2	3.60
NC25141		NC272533	B	2.82
NC25143		NC273530	B	3.01
NC25144		NC275532	C1	3.31
NC25145		NC276533	C1	3.46
NC2516		NC234538	C2	3.59
NC25161	Lochain Dubha	NC297553	C1	3.65
NC25186		NC276507	C2	3.51
NC25187		NC278508	C2	3.31
NC25188		NC279509	C2	3.31
NC25189		NC287512	C2	3.72
NC25190		NC285509	C2	3.74
NC25199	Loch a' Gharbh-bhaid Beag	NC266501	C2	3.57
NC25200		NC285511	A	1.54
NC25201		NC280505	B	2.56
NC2521	Lochan Cul na Creige	NC280534	C2	3.59
NC2522	Loch na Claise Carnaich	NC280525	C2	3.37
NC2523	Loch an Eas Ghairbh	NC269525	C2	3.65
NC2526		NC201539	C2	3.33
NC2528	Loch Sgeir a' Chadha	NC235511	C2	3.65
NC2529		NC238508	C2	3.57
NC2530		NC233504	C2	3.51
NC2531	Mathair a' Gharbh Uilt	NC280507	C2	3.60

80

Site code	Site name	Grid Reference	Lake Group	PLEX Score
NC2532		NC238501	C2	4.00
NC2533		NC241501	C2	3.59
NC2534		NC203537	C2	3.31
NC2535		NC204537	C1	3.02
NC2539	Loch na Thull	NC253503	C2	3.60
NC2541	Loch na Caillich	NC250510	C2	3.59
NC2542		NC290552	B	3.21
NC2543		NC225539	C2	3.80
NC2545		NC212540	C2	3.72
NC2556		NC209533	B	3.65
NC2562		NC219540	C2	3.72
NC2602	Loch a' Gheodha Ruaidh	NC247674	C2	4.12
NC2606	Lochan nan Sac	NC239656	C2	3.65
NC2608	Sandwood Loch	NC228643	C2	4.56
NC2609		NC233632	C2	3.87
NC2610	Loch a' Phuill Bhuidhe	NC268632	C2	3.93
NC2611	Loch na Creige Riabhaich	NC288633	C2	3.97
NC2614	Loch a' Mhuilinn	NC206631	C2	4.44
NC2615	Loch na Gainimh	NC204614	C2	4.36
NC2616	Loch Deibheadh	NC225606	C1	3.08
NC2617	Loch Mor a' Chraisg	NC227602	C2	3.77
NC2618		NC236604	C2	3.85
NC2621		NC244675	C2	3.29
NC2623		NC238657	B	3.08
NC2624		NC235661	C2	3.51
NC2625		NC233659	C2	3.51
NC2628	Lochan Beul na Faireachan	NC242648	C1	3.08
NC2629		NC209615	C1	3.65
NC2645		NC232607	B	3.08
NC2649		NC263637	C1	2.98
NC2650		NC264634	C1	3.77
NC2652		NC268634	B	3.33
NC2654		NC269627	A	1.54
NC2701		NC272730	C2	3.57
NC2702		NC276731	C2	4.52
NC2703	Loch na Seamraig	NC280726	C2	3.65
NC2709		NC282723	B	3.08
NC2712		NC284724	B	3.08
NC3001	Loch Eileag	NC308063	C2	4.34
NC3002	Loch Craggie	NC324054	C2	4.64
NC3003	Lochan a' Bhualt	NC331036	C2	4.20
NC3004	Loch Thurnaig	NC398003	A	2.88
NC3005		NC308066	A	2.31
NC3006		NC309065	B	3.23
NC3007		NC315007	C2	3.41
NC3008		NC312001	C2	3.80
NC3012	Loch a' Bhith	NC327005	C2	3.59
NC3019		NC378048	C2	3.37
NC3020	Loch na Claise Moire	NC385051	C2	4.04
NC3023		NC368032	C2	3.54
NC3024	Loch a' Bhrochain	NC370030	C2	3.65
NC3025		NC375029	C2	3.54
NC3027		NC390008	B	2.56
NC3028		NC391008	D	5.38
NC3101	Loch Carn nan Coubhairean	NC343174	D	4.57
NC3102	Dubh Loch Beag	NC325165	D	4.27
NC3103	Loch Sail an Ruathair	NC334148	C2	4.73
NC3104	Loch Ailsh	NC315110	D	4.96
NC3109		NC324162	A	3.27
NC3110		NC325162	A	2.88
NC3201	Loch a' Ghriama	NC391265	C2	4.23
NC3202	Gorm Loch Mor	NC320245	C2	4.65
NC3203	Fionn Loch Mor	NC335235	C2	4.53
NC3204	Fionn Loch Beag	NC340228	C2	4.32
NC3205	Loch an Eircill	NC308275	C2	3.85
NC3271		NC332242	C1	3.65
NC3279		NC310232	A	2.31
NC3280		NC336232	C1	3.65
NC3281		NC336231	C1	3.08
NC3282		NC332233	B	3.08
NC3286		NC334228	C1	3.65
NC3288		NC342224	C1	3.65
NC3301	Loch nan Ealachan	NC300393	C2	4.71
NC3303	Loch Merkland	NC390315	C2	4.75
NC3304	Loch More	NC330372	D	4.73
NC3308	Lochain nan Ealachan	NC322351	D	5.19
NC3309		NC327346	D	5.62
NC3313		NC335332	B	3.08
NC3314	Loch Srath nan Aisinnin	NC322327	C2	4.23
NC3320		NC321353	D	5.96
NC3321		NC330344	C1	4.36
NC3323	Loch Ulbhach Coire	NC379376	C2	4.44
NC3325	Loch Eas na Maoile	NC371350	C2	3.54
NC3372		NC373350	B	3.35
NC3402	Loch Dionard	NC357490	C2	4.03
NC3406	Loch na Tuadh	NC310472	C2	4.18
NC3409	Loch an Easain Uaine	NC325463	C1	4.69
NC3410	An Dubh-loch	NC355460	D	5.00
NC3414	Lochan Ulbha	NC357455	D	4.51
NC3415		NC371454	C1	4.36
NC3416	Lochan na Faoileige	NC327449	A	3.46
NC3417	Loch na Seilge	NC369446	D	4.62
NC3436		NC300420	C1	2.98
NC3452		NC371457	C1	4.17
NC3453		NC372454	C1	3.77

Site code	Site name	Grid Reference	Lake Group	PLEX Score
NC3454		NC372453	C1	4.15
NC3501	Lochan na Glamhaichd	NC340595	C2	3.97
NC3503	Lochan Sgeireach	NC305560	C2	4.02
NC3504		NC307559	C2	3.88
NC3505		NC390542	C2	3.65
NC3506	Lochan Havurn	NC394543	C1	3.46
NC3508		NC308560	C2	3.37
NC3509		NC397541	C2	3.65
NC3510		NC301557	C2	3.54
NC3511		NC304552	C2	3.65
NC3512		NC348596	C2	4.13
NC3513		NC310560	C2	3.31
NC3515		NC310558	C1	3.46
NC3601	Loch Inshore	NC330696	C2	4.45
NC3602	Lochan nam Breac Buidhe	NC336639	C2	3.74
NC3603	Loch Bad an Fheur-Loch	NC338672	C2	3.57
NC3605	Loch Lanlish	NC385683	I	6.43
NC3606	Loch Croispol	NC390680	E	6.79
NC3607	Loch Borralie	NC383672	E	6.77
NC3608	Loch Caladail	NC396667	E	6.32
NC3610		NC346600	C2	3.46
NC3611		NC349604	C2	3.51
NC3612	Loch Airigh na Beinne	NC325665	C2	4.02
NC3630		NC347602	C1	3.08
NC4001		NC450014	C2	4.81
NC4002		NC477013	D	4.70
NC4012	Loch an Rasail	NC475086	C2	3.80
NC4025	Loch na Fuaralaich	NC487065	C2	3.97
NC4039		NC474016	A	2.31
NC4040		NC474019	B	3.08
NC4104		NC415188	B	3.46
NC4131	Lochan a' Choire	NC462133	C2	3.65
NC4147	Loch Sgeireach	NC457112	C1	4.23
NC4201	Loch Fiag	NC450290	C2	3.67
NC4202	Suil a' Ghriama	NC411279	C2	3.46
NC4203	Loch Strath Duchally	NC423268	C2	3.46
NC4204		NC461251	B	3.85
NC4205	Loch an Fheoir	NC495238	C2	3.37
NC4206	Loch an Ulbhaidh	NC493227	C2	3.75
NC4207	Loch Poll a' Phac	NC462283	C2	3.85
NC4208	Loch Eileanach	NC483277	C2	3.65
NC4209		NC466283	B	3.37
NC4210	Loch Camasach	NC474280	C2	3.72
NC4228		NC448279	B	3.37
NC4229		NC451280	A	2.88
NC4274	Loch an Alaskie	NC478267	C2	3.94
NC4283		NC499238	A	2.31
NC4303	Lochan na Creige Riabhach	NC414376	C1	4.23

Site code	Site name	Grid Reference	Lake Group	PLEX Score
NC4304		NC410373	C1	4.46
NC4305	Loch Coire na Saidhe Duibhe	NC449361	D	4.51
NC4306	Loch an Aslaird	NC424367	C2	3.74
NC4307	Loch an t-Seilg	NC417361	C2	4.56
NC4308	Loch an Tuim Bhuidhe	NC408355	C2	3.91
NC4312	An Glas-lLoch	NC496313	C2	3.90
NC4313		NC411367	C1	3.85
NC4314		NC407369	C2	4.69
NC4343		NC451362	C1	3.85
NC4344		NC453361	C1	3.46
NC4405		NC455493	C1	3.54
NC4406		NC457493	C2	3.59
NC4407	Loch a' Ghobha-Dhuibh	NC497497	D	4.36
NC4501	Loch Hope	NC460540	C2	4.18
NC4502		NC461589	C2	3.54
NC4503	Loch a' Choin-bhoirinn	NC455573	C2	3.51
NC4504		NC446571	C2	3.65
NC4505		NC435568	H	6.73
NC4506	Loch na Creige Duibhe	NC444566	C2	3.19
NC4507		NC427546	C2	3.65
NC4508		NC432544	C2	3.74
NC4509		NC436544	C2	3.54
NC4510	Loch Bealach na Sgeulachd	NC425540	C2	3.57
NC4511		NC428540	C2	3.59
NC4512		NC431537	C2	3.65
NC4513		NC421534	C2	4.40
NC4517	Loch Bacach	NC460520	C2	3.65
NC4521	Lochan na Fearna	NC451502	C2	4.07
NC4522	Dubh-loch na Beinne	NC468506	C2	3.85
NC4523		NC485506	D	4.55
NC4524	Loch na Seilg	NC494510	D	4.62
NC4525		NC465521	C2	3.54
NC4530		NC424534	B	2.77
NC4531		NC426536	C2	3.65
NC4532		NC428536	C2	3.65
NC4535		NC430536	C2	3.54
NC4602	Loch Uamh Dhadhaidh	NC454642	D	4.23
NC4604		NC488617	C1	3.85
NC4605	Loch na Cathrach Duibhe	NC460613	C2	4.00
NC4606	Loch a' Choire	NC466612	C1	2.98
NC4607	Loch Ach'an Lochaidh	NC458607	C2	4.41
NC4608	Loch Cragaidh	NC458602	C2	3.97
NC4609		NC465617	A	2.31
NC4610		NC463614	C1	4.23
NC4629	Loch Meadaidh	NC400644	C2	4.06
NC5001	Loch na Caillich	NC516082	C2	3.97
NC5002		NC518078	C2	3.38

Site code	Site name	Grid Reference	Lake Group	PLEX Score
NC5003		NC580061	D	4.83
NC5004		NC529022	A	2.56
NC5005		NC526019	A	2.31
NC5006		NC520013	A	2.95
NC5007		NC514081	A	1.54
NC5101	Loch an Staing	NC523188	C2	3.46
NC5102		NC578158	A	2.88
NC5202		NC553294	C2	3.46
NC5203	Loch Bad an Loch	NC550289	C2	3.81
NC5204	Loch nan Uan	NC566293	D	4.55
NC5205	Loch an Fheoir	NC510259	C2	3.35
NC5206	Loch Gaineamhach	NC514258	C2	3.42
NC5207		NC515248	C1	3.23
NC5208		NC510242	C2	3.65
NC5209		NC504243	C2	3.77
NC5210		NC504240	C2	3.17
NC5211		NC507238	C2	3.24
NC5212		NC508237	B	3.01
NC5213		NC507234	B	3.17
NC5214	Loch Dubh Cul na Capulich	NC528236	A	2.31
NC5215	Loch na Capulich	NC524223	C2	3.65
NC5218	Loch a' Bhealaich	NC599267	C2	4.59
NC5223		NC510260	B	3.08
NC5224		NC510257	B	2.98
NC5225		NC515255	A	2.69
NC5226		NC516252	A	2.31
NC5227		NC515249	A	2.31
NC5228		NC516248	A	2.31
NC5229		NC508244	B	2.69
NC5230		NC505237	A	2.31
NC5231		NC508240	A	1.54
NC5232		NC509239	A	1.54
NC5233		NC508239	C2	3.57
NC5240		NC527224	A	2.88
NC5241	Loch a' Ghiubhais	NC556236	C1	3.85
NC5301		NC556302	C2	3.59
NC5308	Loch Ben Harrald	NC520330	C2	4.09
NC5309	Loch na Glas-choille	NC559305	C2	3.65
NC5310	Loch an Tairbh	NC568312	C2	3.42
NC5401	Loch an Dherue	NC540480	C2	4.48
NC5405		NC546457	C1	3.40
NC5406		NC550459	C1	2.69
NC5407	Loch a' Mhadaidh-ruaidh	NC554460	C2	4.12
NC5409	Loch Haluim	NC558455	C2	4.21
NC5417	Loch Coulside	NC582435	C2	4.42
NC5419	Loch Dionach-caraidh	NC558402	C2	3.96
NC5420	Loch a' Mhoid	NC569410	C2	3.99
NC5421	Loch Staing	NC579405	C2	3.65
NC5422	Loch Eileanach	NC593402	C2	3.38
NC5423	Loch Meadie	NC500405	C2	4.48
NC5424	Loch na Creige Riabhaich	NC584491	A	3.27
NC5425		NC542452	B	2.77
NC5426		NC544453	C1	3.23
NC5427		NC546450	C2	3.42
NC5428		NC560407	C2	3.51
NC5447		NC553457	B	3.00
NC5448		NC554457	A	2.56
NC5449		NC555459	C2	3.65
NC5464		NC556403	C2	3.96
NC5465		NC572410	A	2.56
NC5466		NC574409	A	1.54
NC5467		NC575409	B	2.69
NC5468		NC576401	B	3.00
NC5471		NC594412	B	2.77
NC5473		NC595407	C1	3.27
NC5474		NC596408	C1	3.23
NC5475		NC596404	C1	3.23
NC5476		NC553452	B	2.98
NC5501	Dubh-loch na Creige Riabhaich	NC506500	C1	4.23
NC5505		NC562579	C2	3.80
NC5507	Loch Fhionnaich	NC555563	C2	3.51
NC5510	Loch na h-Airigh Bige	NC549550	C2	3.80
NC5513	Lochan na Cuilce	NC573535	C2	3.65
NC5514	Lochan Hakel	NC570530	C2	4.39
NC5515	Loch Craisg	NC599577	E	5.00
NC5606	Loch nan Aigheann	NC529680	C1	3.94
NC5608	Lochan na Seilg	NC538656	C1	4.23
NC5609	Loch na h-Uamhachd	NC555658	C2	4.07
NC5611	Loch Fada	NC529645	A	2.31
NC5613	Lochan nam Breac Buidhe	NC525644	C2	3.90
NC5614	Loch a' Mhuilinn	NC569609	C2	4.52
NC5615		NC593619	C2	3.85
NC5616		NC539651	C1	3.08
NC5617		NC536661	C1	3.78
NC5618		NC530635	B	3.08
NC5619	Lochan nam Breac Buidhe	NC527642	C2	4.23
NC5624		NC526642	C2	4.23
NC6001	Loch Tigh na Creige	NC616093	C2	4.73
NC6002	Loch Preas nan Sgiathanach	NC680090	C2	4.20
NC6003	Loch Dola	NC606080	E	5.38
NC6004	Loch Craggie	NC625075	C2	4.29
NC6005	Loch Muidhe	NC665052	C2	4.74
NC6006		NC662050	B	4.63
NC6007	Loch Cracail Mor	NC627021	C2	4.71

Site code	Site name	Grid Reference	Lake Group	PLEX Score
NC6008	Loch Aairighe Mhor	NC692022	C2	4.53
NC6009	Lochan Iain Bhuidhe	NC689012	C2	4.11
NC6010	Lochan a' Ghuibhais	NC699011	B	3.33
NC6017	Lochan na Gaoithe	NC661006	C2	4.10
NC6021	Lochan na Faolaig	NC642010	A	2.31
NC6022	Loch na Saobhaidhe	NC652016	C2	4.44
NC6023		NC653012	A	1.54
NC6101	Glas-loch Mor	NC672195	C2	4.46
NC6102	Loch nam Breac Beaga	NC651187	C2	4.17
NC6103	Lochan Dubh Cadhafuaraich	NC681183	C1	3.40
NC6104	Loch a' Mheallain Leith	NC660172	C2	4.09
NC6105	Loch na Fuaralachd	NC604163	C1	4.04
NC6106	Loch Coire na Bruaiche	NC636165	C2	4.66
NC6107	Loch Beag na Fuaralachd	NC609155	B	2.69
NC6108	Loch an t-Slugaite	NC676158	C2	3.89
NC6109	Lochan Sgeireach	NC670151	C2	3.62
NC6110	Loch Beannach	NC685145	C2	4.23
NC6111	Loch Beannach	NC600125	C2	4.10
NC6112		NC607159	A	2.31
NC6113		NC648139	A	1.54
NC6114		NC600134	A	2.31
NC6117	Lochan Sgeireach	NC670155	C2	3.62
NC6201	Loch Choire	NC635288	D	4.47
NC6202	Glas-loch Beag	NC660200	C2	4.64
NC6305	Loch Naver	NC620370	C2	4.80
NC6307	Loch Tarbhaidh	NC637360	C2	3.80
NC6309	Loch Ruigh nan Copag	NC648353	C2	3.31
NC6310	Loch Coire nam Feuran	NC664350	C2	3.61
NC6311		NC657362	B	3.08
NC6312		NC659350	C2	3.41
NC6401	Loch Loyal	NC621475	C2	4.85
NC6409	Loch Syre	NC661446	C2	4.15
NC6412	Gull Loch	NC675433	A	2.95
NC6502	Lochan Dubh	NC624593	C2	3.97
NC6503	Loch Buidhe	NC631595	C2	4.23
NC6504	Loch Cormaic	NC626581	C2	4.20
NC6505	Loch Crocach	NC643592	C2	4.23
NC6506	Loch Dubh Beul na Faire	NC647593	C2	3.74
NC6507	Clar-loch Mor	NC650586	C2	3.65
NC6508	Loch nan Gamhna	NC645580	C2	3.21
NC6511		NC698592	C2	3.90
NC6516	Loch Arbhair	NC674572	C2	4.47
NC6517	Grian-loch Beag	NC642568	C2	4.58
NC6518	Loch nam Breac Buidge	NC650569	C2	4.23
NC6519	Grian-loch Mor	NC645558	C2	4.36
NC6520	Na Caol Lochan	NC689564	C2	3.46
NC6521	Na Caol Lochan	NC692557	C2	3.65
NC6522	Na Caol Lochan	NC693555	B	3.00

Site code	Site name	Grid Reference	Lake Group	PLEX Score
NC6523	Loch Stephan	NC696552	C2	4.23
NC6524	Loch a' Chnoic Ruadh	NC691541	C2	3.37
NC6525	Lochan nan Carn	NC687537	C2	3.85
NC6526	Lochan nan Carn	NC692533	C2	3.75
NC6527	Loch nan Ealachan	NC677519	C2	3.42
NC6528	Loch nan Con-donna	NC674515	C1	3.65
NC6529		NC685519	B	3.46
NC6530	Loch na Moine	NC628518	B	3.02
NC6531	Loch Craggie	NC615520	C2	4.65
NC6532	Loch Slaim	NC624535	D	4.98
NC6533	Lochan na h-Uimheachd	NC609544	B	3.00
NC6534		NC608540	B	2.98
NC6535	Lochan nan Carn	NC685534	C2	3.74
NC6536		NC693525	C2	3.54
NC6537		NC695525	C2	3.21
NC6539	Lochan Dubh	NC648599	B	3.00
NC6540	Clar-loch Beag	NC652584	C1	2.69
NC6569		NC626536	A	1.54
NC6570		NC630536	A	2.95
NC6571	Lochan Leacach	NC688572	C2	3.41
NC6572		NC689569	C1	2.95
NC6574		NC629515	C1	3.65
NC6576	Na Caol Lochan	NC694556	B	3.08
NC6577		NC683540	A	1.54
NC6579		NC691530	B	2.56
NC6585		NC632595	C1	2.98
NC6586		NC652566	C2	3.19
NC6601	Lochan Ruadh	NC636619	C2	3.94
NC6602	Loch Modsaire	NC649615	C2	4.55
NC6609	Loch Skerray	NC663600	C2	4.00
NC6610	Loch a' Chaoruinn	NC667601	C2	4.60
NC6612	Lochan a' Choire	NC684603	B	3.59
NC6613	Loch Chuibhe	NC695601	C2	4.19
NC6614		NC691607	C2	3.54
NC6615	Lochan Drum an Duin	NC693610	E	4.95
NC7001		NC742070	B	4.78
NC7002	Loch Horn	NC797060	C2	4.00
NC7003	Loch Farlary	NC772050	D	4.96
NC7004		NC722036	B	4.70
NC7005	Loch Salachaidh	NC760037	D	5.52
NC7006	Loch Airighe Bheg	NC703024	C2	4.26
NC7007	Loch Lunndaidh	NC785007	C2	4.45
NC7101	Lochan Dubh Cul na h-Amaite	NC758142	C2	4.23
NC7102	Loch Bad na h-Earba	NC765138	C2	4.36
NC7103	Loch na Glaic	NC759132	C2	3.93
NC7104	Lochan Dubh	NC746122	D	5.19
NC7105	Loch Beannach	NC740113	C2	3.82

84

Site code	Site name	Grid Reference	Lake Group	PLEX Score
NC7106	Loch Grudaidh	NC744101	C2	4.55
NC7117		NC737116	C2	3.77
NC7119	Loch a' Chrioslaich	NC784164	B	3.08
NC7120		NC797168	B	3.08
NC7201	Gorm-loch Beag	NC706273	C2	4.34
NC7202	Gorm-loch Mor	NC713233	D	4.36
NC7213		NC714268	A	2.31
NC7217		NC716229	B	3.08
NC7301	Loch Rimsdale, nan Clar & Badanloch	NC759355	D	5.06
NC7303	Loch an Alltan Fhearna	NC750335	C2	4.34
NC7304	Loch Truderscaig	NC711328	C2	4.42
NC7305	Lochan Dubh	NC719329	A	2.31
NC7306	Loch na Gaineimh	NC767304	C2	4.52
NC7314		NC721331	A	1.54
NC7315		NC721327	B	3.08
NC7417	Palm Loch	NC709410	C2	4.02
NC7421	Loch Rosail	NC715401	C2	3.72
NC7470	Loch Gaineimh	NC798430	C2	3.40
NC7471		NC795413	A	1.54
NC7473		NC795412	A	1.54
NC7501	Lochan Duinte	NC716583	C2	4.81
NC7520	Loch nan Laogh	NC758558	C2	3.46
NC7523		NC765553	C2	3.31
NC7524		NC762550	B	2.56
NC7525	Loch Mor na Caorach	NC764545	C2	4.23
NC7527		NC760541	C2	3.54
NC7607		NC734621	C2	4.74
NC7608	Loch Mor	NC720608	C2	4.23
NC7623	Loch Mer	NC704600	C2	4.38
NC7624		NC717608	C2	4.46
NC7625		NC719604	C1	3.46
NC8001	Loch Brora	NC850080	D	5.29
NC8002	Loch an Tubairnaich	NC875089	E	5.00
NC8003	an Dubh-lochan	NC880077	C1	4.51
NC8201	Loch Ascaig	NC850255	C2	4.66
NC8301	Loch Lucy	NC878394	C2	3.54
NC8302	Loch Culaidh	NC863390	C2	3.97
NC8303	Loch Achnamoine	NC814320	D	5.10
NC8316	Loch Arichlinie	NC847350	C2	5.09
NC8319		NC863352	B	3.08
NC8401	Loch na Saobhaidhe	NC800471	C2	4.16
NC8405	The Cross Lochs	NC865471	C1	4.81
NC8406	The Cross Lochs	NC869469	C2	3.37
NC8407	The Cross Lochs	NC868467	B	3.08
NC8408	The Cross Lochs	NC872464	C2	4.23
NC8409	The Cross Lochs	NC871461	C2	4.40
NC8410	Loch Druim a' Chliabhain	NC810410	C2	4.60
NC8411	Loch Crocach	NC805437	C2	3.96
NC8414		NC875463	B	3.08
NC8416	Loch Coire nam Mang	NC800405	C2	4.47
NC8434		NC805432	A	2.95
NC8435		NC806433	A	2.95
NC8436		NC806431	A	2.95
NC8437		NC808433	A	2.95
NC8440		NC801411	A	1.54
NC8441		NC818421	A	1.54
NC8511	Lochan na Ceardaidh	NC838550	C1	3.65
NC8513	Loch na Main	NC819544	B	3.08
NC8514	Loch Crasgach	NC835543	B	3.92
NC8539	Loch a' Bhroillich	NC814538	C2	4.23
NC8540		NC818539	A	2.31
NC8542		NC819536	C1	3.19
NC8610	Loch nam Breac Beag	NC815606	C2	4.41
NC8611	Loch nam Breac Mor	NC811603	C2	4.23
NC8613	Loch Baligill	NC856620	D	4.76
NC8614	Achridigill Loch	NC858612	C2	4.23
NC8615	Loch Sgiathanach	NC870623	C1	3.78
NC8616	Loch Coulbackie	NC873614	B	3.72
NC8617	Loch Mor	NC889633	D	5.38
NC8618	Loch Earacha	NC898607	C2	3.90
NC8619		NC893613	C1	5.25
NC8620		NC888638	D	4.95
NC8637		NC863613	A	1.54
NC8646		NC892614	C1	5.25
NC9304	Loch Glutt	NC992373	C1	3.69
NC9401		NC926497	C1	4.01
NC9402		NC932496	C1	2.98
NC9403	Lochan nan Clach Geala	NC935494	C1	3.85
NC9408	Loch Sletill	NC958471	C2	4.62
NC9410	Loch Leir	NC955458	C2	4.51
NC9415		NC928498	D	5.00
NC9425	Lochan Ealach Mor	NC966482	C2	4.78
NC9435	Loch na Cloiche	NC975474	C2	4.51
NC9503	Loch na Seilge	NC922586	D	4.73
NC9504	Loch na Caorach	NC913587	C2	4.23
NC9508		NC923578	A	1.54
NC9518		NC909582	A	3.72
NC9519		NC907586	C1	3.81
NC9522	Loch Torr na Ceardaich	NC970510	D	4.48
NC9524	Lochan Dubh Cul na Beinne	NC984544	D	5.34
NC9532		NC974520	D	4.71
NC9535	Loch Tuim Ghlais	NC979525	D	5.19
NC9601		NC903647	C1	4.23
NC9602		NC912645	C1	4.23
NC9605	Caol-loch	NC922616	B	3.08

Site code	Site name	Grid Reference	Lake Group	PLEX Score
NC9606	Loch Akran	NC923605	C2	4.58
NC9607	Loch na Moine	NC937655	C1	4.23
NC9609	Loch Achbuiligan	NC988655	E	6.60
ND0403	Lochan Dubh nan Geodh	ND060478	A	2.31
ND0410	Lochan Chairn Leith	ND057450	C1	3.85
ND0420	Loch Meadie	ND090481	C2	4.87
ND0421	Loch More	ND079455	D	5.87
ND0422	Caol Loch	ND025486	C2	4.81
ND0426	Loch Sand	ND096410	C2	4.87
ND0427	Loch Eileanach	ND070475	C2	4.09
ND0428	Loch Gaineimh	ND050470	E	5.05
ND0501	Loch an Duine	ND044507	B	2.56
ND0506	Loch Losgann	ND026500	C1	3.23
ND0513	Lochan na Saighe Glaise	ND012514	C2	3.65
ND0516	Loch a' Chiteadh	ND040503	D	4.67
ND0520	Loch Caluim	ND022513	D	4.81
ND0521	Loch Scye	ND006554	D	4.28
ND0522	Loch Olginey	ND089576	E	6.19
ND0523	Loch Shurrey	ND045555	D	5.63
ND0611	Lochan Ealach	ND041602	B	4.08
ND0614	Loch Saorach	ND015605	C2	4.23
ND0619	Loch Calder	ND072603	D	5.91
ND0626	Loch Thormaid	ND010605	C2	4.62
ND0702		ND087707	H	6.15
ND0703		ND096711	B	6.09
ND0704		ND096712	C2	4.42
ND0705	Coghill Loch	ND092709	E	5.37
ND0706		ND088710	E	5.83
ND0707	Scrabster Loch	ND087704	E	6.32
ND1306	Lochan Coire na Beinne	ND147399	C1	4.23
ND1402	Loch a' Cherigal	ND100487	E	5.69
ND1403	Loch Rangag	ND178415	E	5.53
ND1404	Loch Ruard	ND142432	C2	3.85
ND1405	Loch Stemster	ND189424	E	5.35
ND1406	Loch Thulachan	ND105412	C2	3.90
ND1513		ND176571	E	6.16
ND1514	Loch Scarmclate	ND189596	I	6.94
ND1515	Loch of Toftingall	ND190523	E	6.32
ND1702	Many Lochs	ND191746	C1	3.27
ND1704	Sanders Loch	ND136747	C1	3.85
ND1706	Many Lochs	ND195747	C1	3.92
ND2412	Loch Camster	ND264442	D	5.93
ND2506	Loch of Winless	ND292547	D	5.61
ND2507	Loch Watten	ND230560	I	7.08
ND2605	Loch Heilen	ND255684	E	6.31
ND2702	Loch Burifa	ND201761	A	2.31
ND2703	Loch of Easter Head	ND207763	D	5.08
ND2704	Many Lochs	ND203750	C1	3.92
ND2705	Long Loch	ND204760	D	4.69
ND2706	Loch of Muirs	ND202734	A	3.72
ND2707	Many Lochs	ND200748	C1	3.92
ND2711	Sanders Loch	ND210754	D	6.15
ND2712	Loch of Mey	ND271736	E	6.09
ND2717	Black Loch	ND203745	C1	3.27
ND2718	St John's Loch	ND225723	E	6.41
ND2801	Loch of Torness	ND254886	B	4.62
ND2802	Loomi Shuns	ND257898	C1	4.23
ND2803	Loomi Shuns	ND258898	C1	3.65
ND2804	Loomi Shuns	ND258899	C1	5.00
ND2901	Berry Lochs	ND242907	B	3.08
ND2903	Heldale Water	ND256924	C1	4.94
ND2904		ND299954	B	3.46
ND2905	Sands Water	ND247939	A	1.54
ND2906	Water of Hoy	ND281999	A	1.54
ND29112	Berry Lochs	ND244906	B	3.08
ND3404	Groat's Loch	ND314408	B	3.78
ND3405		ND301440	E	5.00
ND3406	Loch Hempriggs	ND343470	D	5.88
ND3407	Loch of Warehouse	ND300424	D	5.15
ND3408	Loch of Yarrows	ND310438	E	5.51
ND3409	Loch Sarclet	ND342428	E	6.64
ND3410	Loch Watenan	ND318412	E	5.45
ND3505	Loch of Killimster	ND308560	B	4.87
ND3509	Loch of Wester	ND325593	E	6.87
ND3604		ND354651	D	4.97
ND3608	Loch of Auckengill	ND351652	C2	4.40
ND4801	Trena Loch	ND466852	H	7.46
ND4802		ND469855	G	6.73
ND4803		ND467854	I	6.89
ND4804	Graemston Loch	ND451846	I	7.07
ND4805	Liddel Loch	ND453834	I	7.44
ND4806	Loch of Lythe	ND444858	I	7.20
ND4807		ND462835	G	6.28
ND4808		ND463834	D	6.40
ND4809		ND466835	D	5.93
ND4810		ND464836	I	7.64
ND4901	Echna Loch	ND474967	E	6.39
ND4902		ND484984	I	7.12
ND4903		ND484959	I	7.58
ND4904		ND454913	H	6.54
ND4905		ND436908	H	7.12
ND4906		ND426938	J	8.46
ND4907		ND416938	I	7.97
ND4913		ND428938	I	7.88
ND4914		ND436909	H	7.12
ND4915		ND437907	H	7.12

Site code	Site name	Grid Reference	Lake Group	PLEX Score
ND4916		ND451912	H	6.54
ND4917		ND452913	H	6.54
NF6001		NF688048	B	3.68
NF6002		NF681045	C2	4.42
NF6003		NF685042	C2	3.41
NF6004		NF641003	D	5.49
NF6008	Lochan na Cartach	NF694027	C2	3.42
NF6701		NF688705	E	7.12
NF6702		NF691704	G	7.34
NF7005		NF716013	E	5.62
NF7006		NF710007	C2	4.13
NF7109	Loch an Eilean	NF749189	E	5.85
NF7111	Loch Trosaraidh	NF759170	C2	4.79
NF7112	Loch nan Capull	NF754161	C2	4.62
NF7118		NF746149	G	7.73
NF7201	Loch Ardvule	NF713299	I	7.56
NF7203	Loch Bornish	NF733293	E	6.00
NF7208	Upper Loch Kildonan	NF736280	C2	5.81
NF72105	Loch nam Faoileann	NF751207	E	5.66
NF72150		NF740238	B	4.04
NF72154		NF796246	B	3.02
NF72155		NF757206	D	5.96
NF7222	Loch Aird an Sgairbh	NF733269	C2	5.46
NF7224	Loch Eilean an Staoir	NF733260	E	6.71
NF7240		NF796292	A	1.54
NF7241	Loch nam Faoileann	NF797290	C2	3.68
NF7252	Loch na Liana Moire	NF734249	E	6.47
NF7260	Loch na Cuithe Moire	NF738235	C2	5.30
NF7297		NF739230	E	5.34
NF7299	Loch Hallan	NF736224	E	6.80
NF7303	Grogarry Loch	NF762393	E	6.33
NF73105	Mid Loch Ollay	NF756316	E	6.06
NF73110	Loch Toronish	NF737305	E	6.24
NF7312	Loch Stilligarry	NF763382	I	6.90
NF73125	East Loch Ollay	NF768313	C2	5.02
NF73138	Loch Ceann a' Bhaigh	NF770303	E	6.62
NF7329	Loch an Eilean	NF762372	E	6.26
NF7333	Schoolhouse Loch	NF764366	C2	4.76
NF7334	Loch Eiliean Ghille Ruaidh	NF770367	C2	4.78
NF7340	Loch a' Phuirt Ruaidh	NF768358	C2	4.77
NF7342	Loch Roag	NF756354	E	6.40
NF7344	Loch a' Chnoic Bhuidhe	NF767352	C2	4.54
NF7365	Loch a' Mhoil	NF736348	C2	5.47
NF7366	Loch Altabrug	NF743345	E	5.78
NF7371	Loch Fada	NF756346	C2	5.69
NF7375	West Loch Ollay	NF740327	I	7.10
NF7404		NF761470	I	7.69
NF7477	Loch nam Balgan	NF762400	E	6.67

Site code	Site name	Grid Reference	Lake Group	PLEX Score
NF7507		NF772548	J	8.17
NF7512	Loch Bail 'Fhionnlaidh	NF775537	E	5.96
NF7515		NF762532	E	6.40
NF7537	Loch Mor	NF771524	E	5.85
NF7550	Loch Fada	NF772518	I	7.11
NF7551		NF770510	I	7.62
NF7662		NF780620	G	8.17
NF7667		NF786619	I	7.74
NF7668		NF788623	B	6.04
NF7669	Loch Mor	NF791621	I	7.61
NF7716		NF784759	J	7.69
NF7717		NF784758	J	7.69
NF7723	Loch Grogary	NF716712	E	6.73
NF7725		NF709713	E	6.80
NF7726	Loch nam Feithean	NF712704	I	6.93
NF8106	Loch Marulaig	NF814163	C2	4.19
NF8378	Loch Spotal	NF833366	C2	3.80
NF85197	Loch Hermidale	NF827525	C2	4.29
NF8520	Loch an Fhaing	NF845576	I	6.89
NF8578		NF844555	D	4.55
NF8586	Loch na Deighe Fo Dheas	NF833530	C2	4.21
NF86416	Loch na Cleibh	NF877606	C2	3.54
NF86419		NF882607	C2	3.41
NF86485	Loch Scadavay	NF878663	C2	3.57
NF8779	Loch na Morgha	NF870743	C2	5.15
NF8780	Loch Steinavat	NF875742	E	5.36
NF9690	Loch Surtavat	NF912607	C2	3.77
NF9691		NF957697	B	3.03
NF9721		NF910724	D	4.81
NF9722		NF912723	F	4.81
NF9723		NF913724	B	3.27
NF9738		NF914715	D	4.66
NF9739		NF915714	C1	4.23
NF9740		NF918716	B	2.95
NF9790		NF989740	D	4.36
NF9791		NF988737	C1	3.91
NF9792		NF988736	C2	4.37
NF9793		NF990736	C2	3.94
NF9794		NF989736	B	5.00
NF9798		NF971725	B	3.37
NF9802	Loch Bhruist	NF917825	I	7.46
NF9803	Little Loch Borve	NF913817	I	7.23
NF9805	Loch Watersee	NF931818	H	7.33
NF9806		NF974834	D	4.87
NF9807	Loch a' Mhachaire	NF971839	H	7.01
NF9901		NF998997	C1	3.52
NF9902		NF993994	C1	3.52
NF9903		NF993993	B	3.00

Site code	Site name	Grid Reference	Lake Group	PLEX Score	Site code	Site name	Grid Reference	Lake Group	PLEX Score
NG0801		NG004887	C2	3.02	NG3207		NG333272	B	2.82
NG0804	Loch Steisevat	NG015875	C2	4.98	NG3208	Loch a Bhac-ghlais	NG341278	C2	4.30
NG0806	Loch na Moracha	NG022884	C2	4.34	NG3209		NG342273	C1	4.23
NG0813		NG045835	D	5.76	NG3210		NG343272	C1	4.23
NG08137		NG067871	C2	3.38	NG3211	Loch Bioda Mor	NG370277	C1	3.52
NG08248		NG058877	C2	3.74	NG3212		NG347267	C2	4.23
NG0902		NG024993	J	7.69	NG3301	Loch Duagrich	NG397397	E	5.52
NG0914	Loch Cistavat	NG034955	E	6.12	NG3304		NG311329	C2	3.52
NG0930		NG093972	C2	3.65	NG3305		NG316328	C2	3.32
NG1402	Loch Mor	NG144448	E	6.15	NG3306		NG314327	C2	3.46
NG1404	Loch Eishort	NG161458	C2	4.42	NG3307		NG313327	A	2.88
NG1801		NG104894	C2	3.81	NG3308	Loch Dubh	NG315328	C2	3.85
NG1804			J	7.31	NG3309	Loch a' Ghille-chnapain	NG330329	C2	4.38
NG19107	Loch nam Uidhean	NG145958	C2	3.92	NG3310	Loch nan Uan	NG331327	C2	4.12
NG19252		NG139900	C1	3.69	NG3315	Loch Lic-aird	NG333320	C2	3.38
NG1974		NG105961	B	3.37	NG3316		NG337321	C2	4.08
NG1975	Loch Laxdale	NG108962	C2	3.81	NG3317	Loch Fada	NG340317	C2	4.69
NG1976		NG108950	C2	3.59	NG3318		NG343318	C1	3.94
NG2401		NG206435	C1	4.10	NG3319		NG344319	A	2.95
NG2402		NG261466	A	1.54	NG3322		NG344312	B	2.69
NG2403		NG268462	B	3.02	NG3401	Loch Niarsco	NG391472	D	5.24
NG2503	Loch Vorvin	NG287547	C2	4.23	NG3402	Loch Ravag	NG380450	C2	4.65
NG2504		NG200520	E	5.38	NG3404	Loch Connan	NG387430	D	5.27
NG2505	Loch Corlarach	NG234520	C2	4.64	NG3504		NG373555	B	4.93
NG2506	Loch Suardal	NG240510	D	4.97	NG4006	Loch Iain	NG405007	C1	3.85
NG2601		NG231605	C2	5.05	NG4101		NG438191	B	2.56
NG2901		NG205988	C2	3.46	NG4102		NG441189	C2	3.21
NG2917		NG233950	C1	4.17	NG4106		NG433184	C2	3.32
NG2918		NG234951	C2	3.59	NG4107		NG433182	C1	3.01
NG2920		NG241953	C1	3.97	NG4108		NG436182	B	2.77
NG2922		NG240948	C1	3.61	NG4109		NG443183	C1	3.01
NG2923		NG241950	C1	3.68	NG4120		NG428178	C1	2.98
NG2924		NG243949	C1	4.18	NG4121		NG428177	C2	3.37
NG2925		NG243946	C1	3.40	NG4122		NG432180	C2	3.17
NG2926		NG232943	D	5.38	NG4123	Loch Meachdannach	NG438179	C2	3.46
NG2927		NG232947	C2	4.15	NG4201		NG474297	B	2.86
NG3003		NG253031	C2	3.46	NG4202		NG475295	B	3.17
NG3004	Loch Sgaorishal	NG346022	C2	3.59	NG4203		NG474289	B	2.86
NG3006		NG358017	C2	3.77	NG4204		NG480289	C2	3.29
NG3017		NG358012	B	3.46	NG4205	Loch Dubh	NG484281	C2	3.41
NG3025		NG386016	C2	3.37	NG4207		NG480271	B	3.01
NG3028		NG392013	C1	3.46	NG4208		NG481272	C1	2.98
NG3031		NG393012	C1	3.46	NG4209	Loch a' Choire Riabhaich	NG479265	C1	3.21
NG3201	Loch Sleadale	NG345290	C2	4.28	NG4210		NG480263	A	2.56
NG3202		NG353293	C2	3.35	NG4211		NG468257	A	2.31
NG3204		NG345285	A	2.69	NG4212	Lochan Dubha	NG496243	C2	3.41
NG3205		NG334276	C2	3.46	NG4213		NG498240	B	3.37
NG3206		NG330274	A	2.31	NG4214	Loch an Fhir-bhallaich	NG429208	C2	3.21

Site code	Site name	Grid Reference	Lake Group	PLEX Score
NG4215		NG444209	A	2.31
NG4217	Loch Coruisk	NG482205	C2	3.65
NG4218	Loch a' Choire Riabhaich	NG496210	C2	3.57
NG4219	Loch Coir a' Ghrunnda	NG451202	B	3.08
NG4221		NG423201	C1	3.01
NG4222		NG490278	B	3.08
NG4223		NG472297	B	3.02
NG4224		NG476295	B	3.08
NG4303		NG420347	C1	4.45
NG4305		NG432338	C2	4.23
NG4306		NG429335	C2	4.33
NG4307		NG426332	D	4.81
NG4309		NG468322	A	2.69
NG4310		NG468321	A	2.31
NG4311		NG469323	B	3.22
NG4312		NG468320	A	2.69
NG4313		NG468319	B	3.24
NG4314	Loch a' Sgath	NG471319	C2	3.19
NG4315		NG472317	B	2.56
NG4316		NG473316	B	3.46
NG4317		NG474315	B	2.56
NG4318		NG474316	B	3.08
NG4320		NG472312	B	3.24
NG4321		NG477314	B	3.08
NG4322	Loch Mor na Caiplaich	NG475311	C2	3.54
NG4323		NG477310	B	3.08
NG4324		NG479309	B	3.40
NG4325	Loch nan Eilean	NG471305	B	3.15
NG4326		NG477306	B	3.16
NG4327		NG470303	C2	3.88
NG4329	Loch Caol	NG475300	C2	3.79
NG4330		NG474314	B	3.46
NG4331		NG472301	A	1.54
NG4401	Loch Fada	NG493494	E	6.02
NG4403		NG410446	B	3.21
NG4404	Loch a Ghlinne Bhig	NG412444	C1	3.52
NG4501	Loch Beag	NG401594	C2	3.65
NG4502		NG473597	B	3.56
NG4503		NG473595	B	4.42
NG4504	Loch Cuithir	NG475596	C1	4.62
NG4505	Loch Liuravay	NG485588	E	5.24
NG4506		NG487585	C2	3.46
NG4601		NG402698	B	3.69
NG4602	Loch Sneosdal	NG414692	E	5.45
NG4603		NG457695	B	3.85
NG4604	Loch Fada	NG458696	C2	4.23
NG4605		NG464691	C2	3.37
NG4606		NG465695	C2	4.46

Site code	Site name	Grid Reference	Lake Group	PLEX Score
NG4607		NG470697	C1	4.55
NG4608	Loch Leum na Luirginn	NG447676	C2	4.44
NG4609	Loch Cleat	NG447672	C2	4.23
NG4610	Loch Cleap	NG467662	C2	5.07
NG4611		NG462648	A	3.85
NG4612	Loch Corcasgil	NG452642	C2	3.94
NG4613	Loch Dubhar-sgoth	NG457640	D	5.00
NG4614	Feur Lochan	NG408608	B	4.23
NG4615	Loch Mor	NG405603	C2	3.77
NG4616	Loch Sheanta	NG471698	D	5.38
NG4617		NG407607	A	2.31
NG4701		NG416742	C2	5.08
NG4702	Loch Droighinn	NG455712	C1	3.40
NG4703	Loch Langaig	NG467707	C2	3.73
NG4704		NG467706	C1	3.85
NG4705	Loch Hasco	NG456701	C2	4.07
NG4706	Loch Leum nam Bradh	NG469700	D	4.81
NG4707		NG464710	C1	4.36
NG5001		NG594072	C2	3.21
NG5002	Loch Nigheann Fhionnlaidh	NG583066	C2	4.31
NG5003	Loch a' Ghlinne	NG596054	C2	4.81
NG5013	Loch Aruisg	NG574009	B	4.48
NG5102		NG559194	B	2.69
NG5103		NG556189	B	3.40
NG5104	Loch nan Learg	NG595191	C2	3.42
NG5105		NG544177	C2	4.47
NG5106	Reservoir	NG530154	B	4.49
NG5107		NG598168	C2	3.19
NG5108	Loch an Leoid	NG599163	C2	4.69
NG5109	Loch Gauscavaig	NG591106	C2	4.40
NG5201	Loch nam Madadh Uisge	NG568272	B	3.03
NG5202	Lochain Stratha Mhoir	NG564254	C2	3.51
NG5203		NG561247	C2	3.57
NG5204	Loch an Athain	NG511226	C2	3.97
NG5205	Loch na Sguabaidh	NG560233	C2	3.92
NG5206	Loch na Creitheach	NG514205	C2	4.03
NG5207	Loch Fionna choire	NG538211	B	4.23
NG5208		NG542296	A	1.54
NG5209		NG550291	C1	3.65
NG5211		NG568201	C1	4.56
NG5301	Loch na Meilich	NG574398	D	4.97
NG5302		NG570396	C2	3.41
NG5303		NG556392	B	2.69
NG5304		NG575393	A	2.95
NG5305		NG578393	C1	3.40
NG5306	Loch a Chadha charnaich	NG585392	C2	4.89
NG5307		NG553385	C2	3.65
NG5308		NG553384	C2	3.74

Site code	Site name	Grid Reference	Lake Group	PLEX Score
NG5309		NG553383	C2	3.55
NG5310		NG554380	C2	4.07
NG5311	Loch Storab	NG565386	C2	3.19
NG5312		NG569389	B	3.08
NG5313		NG571383	C2	3.65
NG5314		NG574388	C2	3.57
NG5315	Loch na Mna	NG579387	D	5.14
NG5316	Loch Fada	NG502375	C2	3.42
NG5317		NG554373	C2	3.54
NG5318		NG554372	B	2.77
NG5319		NG559376	C2	3.41
NG5320		NG560374	C2	3.27
NG5321	Loch a' Mhuilinn	NG552367	B	4.23
NG5322		NG563356	C2	3.85
NG5323		NG556309	C2	3.37
NG5326		NG580388	C1	3.08
NG5327		NG578397	A	2.88
NG5402	Loch Mallaichte	NG597494	B	2.69
NG5403		NG599492	C2	3.94
NG5404		NG599490	C2	3.67
NG5405		NG572478	B	3.02
NG5406	Loch na Cuilce	NG573474	B	3.65
NG5407	Loch na Uachdair	NG583470	C2	4.13
NG5408	Loch Beag	NG584472	C2	3.65
NG5409	Loch na Bronn	NG576465	C2	3.85
NG5410		NG581461	B	2.82
NG5411		NG552416	B	4.09
NG5412	Loch an Raithaid	NG553415	E	5.66
NG5413	Loch Eadar da Bhaile	NG557407	C2	4.20
NG5414	Loch Meall Daimh	NG575401	D	4.46
NG5501	Loch Scamadal	NG501549	C1	3.21
NG5502		NG502541	C1	2.98
NG5504	Loch Leathan/Storr Lochs	NG505515	D	5.97
NG5506	Loch Mor	NG588512	C2	4.01
NG5507		NG590512	C2	3.19
NG5508		NG591511	C2	3.46
NG5601	Loch Mealt	NG505650	E	5.36
NG6008	Loch Barabhaig	NG684098	C2	3.94
NG6010	Loch Dhughaill	NG614082	C2	4.29
NG6012	Loch nan Uamh	NG633083	C2	4.41
NG6013	Loch Ic Iain	NG601070	C2	4.58
NG6101	Loch Lonachan	NG627192	C2	4.93
NG6102		NG639195	B	4.17
NG6103	Loch Buidhe	NG639193	C2	5.38
NG6104		NG638192	C2	5.05
NG6105		NG637192	C2	4.62
NG6106	Loch an Starsaich	NG646190	C2	5.02
NG6107		NG650193	B	3.08

Site code	Site name	Grid Reference	Lake Group	PLEX Score
NG6108		NG657195	C2	3.57
NG6109		NG669193	B	2.98
NG6110	Loch Dubh nan Breac	NG671194	C2	3.59
NG6111		NG673192	A	1.54
NG6113		NG682197	B	2.98
NG6114	Loch an Eilein	NG640186	C2	5.15
NG6117		NG604170	B	4.06
NG6118	Loch Fada	NG603168	C2	3.85
NG6119		NG601162	C2	3.65
NG6122	Lochan Iasgaidh	NG673142	C2	4.09
NG6139		NG654111	C2	3.46
NG6140	Loch Meodal	NG657111	C2	4.56
NG6141		NG665110	A	4.04
NG6142	Loch nan Dubhrachan	NG675105	C2	4.23
NG6143	Loch an Ime	NG679102	B	3.02
NG6214	Loch Ashik	NG690232	C2	3.57
NG6215		NG684221	B	2.69
NG6216	Lochain Teanna	NG683219	C2	3.32
NG6217	Lochan Cruinn	NG680217	C2	3.32
NG6218		NG693220	B	3.40
NG6219		NG698216	A	2.31
NG6220		NG677212	A	1.54
NG6221	Loch Cill Chriosd	NG611204	C2	4.81
NG6222	Loch an Droma Bhain	NG667200	C2	4.10
NG6223	Lochain Dubha	NG675206	C2	3.96
NG6224		NG679207	C2	3.76
NG6225		NG678204	C2	4.06
NG6226		NG677202	B	2.69
NG6227		NG680200	A	1.54
NG6228	Loch Airigh na Saorach	NG682202	C2	3.78
NG6229	Lochain a' Mhullaidh	NG689208	C1	3.46
NG6230	Lochain a' Mhullaidh	NG690209	B	2.77
NG6401		NG601489	C1	2.98
NG7105	Loch nan Uranan	NG794165	C2	3.85
NG7106		NG783149	B	4.06
NG7201		NG775290	B	3.02
NG7202	Loch Cul Duibh	NG777288	A	2.31
NG7203	Loch Scalpaidh	NG780287	C2	4.35
NG7204	Loch Palascaig	NG788292	B	4.21
NG7205	Loch Iain Oig	NG792290	C2	4.49
NG7206		NG787285	A	2.95
NG7207		NG718255	B	3.00
NG7208		NG730250	C1	4.18
NG7209		NG723242	B	3.00
NG7212		NG725233	C2	3.65
NG7213		NG730231	C2	3.97
NG7214	Lochan na Saile	NG726228	C2	4.23
NG7215		NG729227	C2	4.02

Site code	Site name	Grid Reference	Lake Group	PLEX Score
NG7301	Loch Erbusaig	NG770303	B	3.86
NG7338		NG741374	C2	3.85
NG7339	Lochan an t-Sagairt	NG743375	C2	3.46
NG7352		NG735364	C2	4.04
NG7358		NG738368	C2	3.97
NG7359		NG741366	C2	3.21
NG7364	Loch Airigh Alasdair	NG744369	C2	3.74
NG74114		NG742426	C2	3.65
NG74117	Lochan Leathann	NG744425	C2	3.54
NG74119	Loch a' Chaorainn	NG748425	C2	4.23
NG7454	Loch a' Mhuilinn	NG708436	C2	4.27
NG7466		NG734432	C1	3.65
NG7467	Lochan Sgeirach	NG736433	C1	3.46
NG7470		NG738434	C1	3.40
NG7474	Loch Odhar	NG751432	C1	4.15
NG7490		NG730425	C1	3.94
NG7493	Lochan na Teanga	NG732425	C2	4.23
NG7527	Loch na Caorach	NG747566	C2	4.23
NG7530	Lochan Fada	NG734558	C2	3.94
NG7531	Lochan Fada	NG736558	C2	3.65
NG7532		NG737555	C2	3.59
NG7534		NG746558	C2	3.31
NG7536	Loch a' Choire Bhuidhe	NG754555	C2	4.52
NG7537	Loch na Creige	NG768556	C2	4.12
NG7606		NG731680	C2	3.21
NG7608		NG740677	B	3.01
NG7609		NG740675	C1	3.00
NG7610		NG742676	A	2.56
NG7710	Loch Bad na h-Achlaise	NG770735	D	5.04
NG7711	Loch nam Breac Odhar	NG766720	C2	4.23
NG7713	Loch Bad a' Chrotha	NG787728	C2	4.59
NG7715		NG767717	C2	3.90
NG7716	Loch Clair	NG773720	C2	4.32
NG7722	Loch Braigh Horrisdale	NG798705	C2	4.50
NG7723	Lochan Fuar	NG798710	C2	4.47
NG7802	Loch Ceann a' Charnaich	NG778895	C2	4.52
NG7811	Loch na Feithe Dirich	NG787887	C2	4.94
NG7822	Loch an t-Seana-bhaile	NG763809	C2	4.23
NG7914	Loch an Draing	NG775905	C2	4.46
NG7915		NG780900	C2	3.41
NG8108	Loch na Lochain	NG810132	D	4.95
NG8201		NG804297	C2	4.35
NG8202	Loch a' Bhealaich	NG830294	C2	4.49
NG8203	Loch a' Ghlinne Dhuirch	NG836298	C2	3.88
NG8204		NG854291	D	5.36
NG8301		NG829327	C2	3.97
NG8302	Lochan Dubha	NG832328	C2	3.59
NG8303		NG835327	B	3.68

Site code	Site name	Grid Reference	Lake Group	PLEX Score
NG8304		NG833325	C2	3.65
NG8309	Loch Lundie	NG806317	C2	3.85
NG8310	Loch Achaidh na h-Inich	NG812307	C2	4.67
NG8311		NG834321	C2	4.51
NG8312	Loch nan Gillean	NG839323	C2	3.92
NG8313	Loch na Leitire	NG844322	C2	3.88
NG8314		NG839321	C2	3.85
NG8315		NG835312	C2	3.51
NG8317	Loch na Doire Moire	NG827304	C2	3.80
NG8318		NG829307	C2	3.21
NG8319		NG832306	B	3.29
NG8320		NG832305	B	3.78
NG8321	Loch nam Breac Mora	NG837308	C2	4.27
NG8322	Loch na Smeoraich	NG843304	C2	4.02
NG8325		NG844325	C2	3.76
NG8330	Lochan Dubha	NG833372	B	3.55
NG8332		NG828368	C1	3.65
NG8333	Loch an Eich-usige	NG835368	B	3.90
NG8440	Loch Coultrie/Loch an Loin	NG855455	C2	4.51
NG8501	Loch Diabaigas Airde	NG816596	C2	4.38
NG8526	Loch Dughaill	NG827514	C2	3.97
NG8653	Loch a' Chaorainn Beag	NG802617	C2	4.12
NG8654	Loch na Feannaig	NG804614	B	3.51
NG8655	Lochan Dubh	NG812615	C2	3.80
NG8656	Loch Freumhach	NG814617	C2	3.85
NG8674	Loch a' Mhuillaich	NG807601	D	4.73
NG87109		NG810702	C2	4.23
NG87110	Lochan Sgeireach	NG810707	C2	3.94
NG8722	Loch Tollaidh	NG840785	C2	3.80
NG8777		NG855741	C2	3.65
NG8783	Loch Doire na h-Airighe	NG872740	C2	3.97
NG8788	Loch nam Buainichean	NG853736	C2	4.34
NG8791		NG862732	C2	4.00
NG8797	Am Feur-Loch	NG859720	C2	3.65
NG8804	Loch Sguod	NG810874	C2	3.65
NG8806	Loch a' Bhaid-luachraich	NG894861	C2	4.14
NG8810	Loch nan Dailthean	NG877830	C2	4.26
NG8816	Loch Chriostina	NG833820	C2	4.00
NG8818		NG872827	C2	4.23
NG8823	Loch na Cloich	NG827817	C2	3.46
NG8947	Loch na Beiste	NG884943	C2	4.20
NG8951	Loch an t-Slagain	NG855938	C2	4.13
NG8952	Loch an Fheoir	NG862933	C2	3.65
NG8953		NG864936	C1	3.00
NG8967	Loch Caol na h-Innse-geamhraidh	NG872929	C2	3.74
NG8981	Loch na h-Innse Gairbhe	NG874939	C2	3.81
NG9008	Loch Coire Shubh	NG961054	C2	3.46

Site code	Site name	Grid Reference	Lake Group	PLEX Score
NG9009		NG964046	C2	3.90
NG9010	Loch Coire nan Cnamh	NG974038	B	3.69
NG9011	Loch a' Choire Bheithe	NG980036	C2	3.59
NG9103	Loch Shiel	NG944183	D	5.02
NG9203		NG987288	B	2.77
NG9204		NG991285	B	3.46
NG9205	Loch nan Ealachan	NG998285	C2	3.88
NG9211		NG940267	C1	2.98
NG9212		NG942264	C2	4.09
NG9213	Loch nan Eun	NG951265	C2	4.23
NG9215		NG948260	C1	4.18
NG9222		NG957261	C2	3.54
NG9224		NG959260	C1	3.27
NG9225		NG956258	C2	3.54
NG9226		NG956257	C2	4.17
NG9227		NG952254	B	2.95
NG9229	Loch an t-Sabhail	NG960256	C2	4.40
NG9230		NG961256	A	2.31
NG9236		NG959253	A	2.56
NG9237		NG951251	C1	3.65
NG9238		NG953250	C1	3.27
NG9254		NG942210	C2	4.23
NG9257		NG951251	C1	3.40
NG9258		NG953261	C1	3.46
NG9329		NG907320	C1	3.21
NG9330		NG910320	C2	3.55
NG9331	Lochan Dubha	NG911318	C2	3.67
NG9418	Loch Coire an Ruadh-Staic	NG920488	D	4.51
NG9433	Loch Dughaill	NG995470	C2	4.78
NG95109	Loch na Craoibhe-caorainn	NG927507	C2	3.80
NG95125	Loch Clair	NG999573	C2	4.68
NG9564	Loch an Uillt-bheithe	NG920523	C2	4.45
NG9587	Lochan Domhain	NG919517	C2	4.41
NG9593	Lochan Eion	NG924511	C2	4.32
NG9631	Loch Bhanamhoir	NG975645	D	4.51
NG9646	Loch Allt an Daraich	NG989633	C1	3.85
NG9647		NG991634	C1	2.95
NG9719	Fionn Loch	NG950785	C2	4.29
NG9735	Loch Maree	NG937718	D	4.77
NG9746		NG921719	C2	4.45
NG9747		NG923720	C2	3.90
NG9748		NG919725	C2	3.81
NG9749		NG917724	B	3.08
NG9864		NG921852	C2	3.59
NG9865		NG922853	C2	3.46
NG9867		NG923854	C2	3.65
NG9871		NG928855	C1	3.21
NG9874		NG931855	C2	3.65
NG9875	Loch nan Eun	NG932852	C2	3.65
NG9876		NG935854	C2	3.32
NG9882	Loch Mhic'ille Riabhaich	NG906845	C2	3.81
NG9893	Lochain Cnapach	NG927847	C2	3.65
NH0001	Loch a' Mhaoil Dhisnich	NH079099	C2	3.74
NH0003	Loch Fearna	NH055031	C2	4.23
NH0106		NH074113	B	4.33
NH0202		NH001284	B	2.82
NH0203		NH004286	C2	3.51
NH0204		NH008287	C1	3.46
NH0205		NH006288	C2	4.00
NH0206		NH003281	A	2.95
NH0207		NH011281	C2	3.55
NH0208		NH015286	B	3.08
NH0209		NH019288	B	3.08
NH0212	Loch na Leitreach	NH020275	C2	4.23
NH0215		NH024266	A	2.31
NH0216		NH025265	B	3.52
NH0217		NH023261	B	3.37
NH0218	Loch Lon Mhurchaidh	NH034264	C2	4.06
NH0219		NH031260	B	4.23
NH0224	Loch Thuill Easaich	NH027232	C2	4.23
NH0229	Loch a' Bhealaich	NH023212	C2	4.23
NH0232	Loch Gaorsiac	NH024214	C2	4.39
NH0301	Loch Calavie	NH051387	C2	4.33
NH0302	Loch Cruoshie	NH055363	C2	4.47
NH0303		NH059365	B	3.27
NH0304		NH060363	C2	4.20
NH0305		NH060365	A	2.56
NH0306		NH061365	A	2.31
NH0307		NH067373	C1	2.95
NH0308		NH068371	A	2.56
NH0309		NH070372	C1	2.98
NH0310		NH072372	C2	3.19
NH0311		NH079370	C2	4.09
NH0314	Loch Goblach	NH084371	C2	3.55
NH0315		NH082376	C2	3.46
NH0316	Loch an Tachdaich	NH093380	C2	4.55
NH0402	Lochan Gaineamhach	NH093451	C1	3.75
NH0502		NH037597	C2	3.94
NH0505		NH040592	A	2.31
NH0506		NH041597	C2	3.94
NH0507		NH041598	C2	4.00
NH0509	Lochain Feith an Leothaid	NH043592	C2	4.01
NH0517	Loch Crann	NH090577	C2	4.08
NH0526		NH042589	C2	3.65
NH0527		NH050586	C2	4.23
NH0528		NH051587	C2	4.32

Site code	Site name	Grid Reference	Lake Group	PLEX Score
NH0530		NH054588	C2	4.23
NH0547	Loch Coulin	NH013553	C2	4.60
NH0719	Lochan Fada	NH025710	C2	4.44
NH0821		NH092813	C1	3.65
NH0824	Lochain Dubh	NH097812	C2	4.04
NH0826		NH099811	C2	4.23
NH0831		NH098810	C1	4.40
NH0832		NH096808	C2	4.04
NH0903		NH042931	C1	3.46
NH0904	Loch na h-Uidhe	NH046933	C2	3.94
NH0905	Loch na Coireig	NH051932	C2	4.07
NH0909		NH051927	C1	3.65
NH0910		NH053928	C2	3.77
NH1009	Lochan Bad an Losguinn	NH158038	C2	3.46
NH1015	Lochan Torr a' Gharbh-uillt	NH167019	B	3.00
NH1016	Loch Poulary	NH124013	C2	4.20
NH1201		NH191232	C2	4.73
NH1202		NH193228	A	2.95
NH1203	Loch Pollain Buidhe	NH189224	B	2.69
NH1205	Loch Affric	NH160222	C2	4.59
NH1206		NH174228	B	3.68
NH1210	Loch Coulavie	NH133215	C2	3.65
NH1211	Loch na Camaig	NH138211	C2	4.23
NH1505		NH161572	A	2.95
NH1507	Loch Gowan	NH153563	C2	4.41
NH1609	Loch na Moine Beag	NH195626	C2	3.85
NH1739	Loch a' Mhadaidh	NH199733	C1	4.33
NH18102		NH146809	C2	3.77
NH18105	Loch nan Eun	NH153817	C2	3.74
NH18107		NH156810	C2	3.62
NH18108		NH159814	C2	4.23
NH18109		NH143807	C2	3.65
NH18110		NH144802	C1	3.77
NH18115		NH154803	C2	3.65
NH1825	Loch Coire Chaorachain	NH106822	C2	4.07
NH1836		NH108812	C2	3.94
NH1899		NH145814	C2	3.85
NH19103	Lochanan a' Mhuilinn	NH193916	C2	3.94
NH19104		NH195916	B	3.23
NH19107		NH196916	B	2.56
NH19108		NH197913	C2	3.65
NH19109		NH193911	C1	3.27
NH2043	Loch Lundie	NH296035	C2	4.11
NH2044	Loch Garry	NH220020	C2	3.93
NH2122		NH202115	C2	3.46
NH2124		NH200114	C2	4.00
NH2201	Coire Loch	NH294282	A	2.69
NH2202	Loch Innis Gheamhraidh	NH290269	C2	3.57

Site code	Site name	Grid Reference	Lake Group	PLEX Score
NH2203	Loch an Amair	NH264260	B	3.37
NH2204	Loch an Gabhlach	NH261256	B	2.69
NH2205	Loch Carn na Glas-leitire	NH255252	C2	3.46
NH2206	Loch an Eang	NH249235	C2	4.07
NH2302	Loch a' Mhuilidh	NH274380	C2	4.31
NH2304	Loch Carrie	NH267333	C2	4.59
NH2307	Loch a' Bhana	NH225314	C2	4.81
NH2308	Loch Sealbhanach	NH238320	C2	4.90
NH2314		NH262330	B	5.63
NH2315		NH255329	D	5.96
NH2415	Lochan Dubh nam Biast	NH287494	C2	3.68
NH2416	Loch Airigh Lochain	NH287486	C2	3.91
NH2507		NH215531	A	2.95
NH2509		NH215530	A	1.54
NH2512	Loch Beannacharain	NH233513	C2	4.10
NH2615	Loch Achanalt	NH274608	C2	4.67
NH2707	Loch a' Gharbhrain	NH281760	C2	4.02
NH2716	Loch Droma	NH260751	D	5.05
NH2807		NH264818	C1	3.65
NH2809	Loch a' Choire Ghranda	NH269805	C1	4.42
NH2958	Loch an Eilean	NH243953	C2	3.80
NH2959		NH246954	C2	3.19
NH2960		NH252955	C2	3.77
NH2963		NH255952	C1	3.23
NH2964		NH257954	C2	3.46
NH2966		NH260952	C2	3.85
NH2978	Loch Coire na Ba Buidhe	NH200918	C1	4.18
NH2979		NH200915	C2	3.40
NH2980	Loch an Acha	NH203914	C2	4.04
NH2981		NH206914	C1	4.23
NH3015	Loch Uanagan	NH369070	C2	5.19
NH3106	Bhlaraidh Reservoir	NH354194	B	3.08
NH3206	Loch Caoireach	NH325272	B	4.09
NH3207	Loch nam Freumh	NH328269	B	3.35
NH3208	Loch na Beinne Moire	NH325265	C2	3.65
NH3211	Loch a' Ghreidlein	NH319260	C2	4.23
NH3224		NH327239	B	3.32
NH3226		NH323238	A	2.31
NH3227		NH321238	B	3.01
NH3236		NH352231	C2	4.41
NH3237		NH352229	C2	3.74
NH3259	Loch ma Stac	NH342215	C1	3.85
NH3262		NH363203	A	2.31
NH3263		NH361204	C1	3.85
NH3264		NH355204	C1	3.94
NH3267	Loch nam Brathain	NH391218	C2	3.46
NH3268		NH394216	B	3.08
NH3271	Loch Liath	NH397209	C2	3.37

Site code	Site name	Grid Reference	Lake Group	PLEX Score
NH3272	Loch na Feannaig	NH396201	C2	3.77
NH3274	Loch an Dubhair	NH391201	C2	3.80
NH3275		NH385200	C1	3.94
NH3276		NH381202	C2	3.94
NH3302	Loch an Airigh Fhraoich	NH335384	B	3.08
NH3312		NH386360	B	4.33
NH3314		NH391351	B	2.69
NH3316		NH396353	A	2.56
NH3317		NH377353	B	4.90
NH3318	Loch Carn nam Badan	NH396345	C2	3.57
NH3322	Lochan na Craoibhe-fearna	NH371329	C2	4.23
NH3323	Lochan Mhairi	NH392324	B	3.46
NH3324	Lochan Dubh	NH370323	B	3.23
NH3325		NH380350	B	4.90
NH3326		NH379352	B	4.90
NH3329		NH386360	B	4.33
NH3401	Loch na Beiste	NH390422	C2	4.07
NH3524	Loch Meig	NH361557	D	4.31
NH3613	Lochan nam Breac	NH380665	C2	3.85
NH3701	Gorm Loch	NH331798	C2	3.80
NH3801		NH330825	A	2.95
NH3802	Lochan Dubh	NH332825	C1	3.27
NH4014	Dubh Lochan	NH446069	C2	3.94
NH4015	Dubh Lochan	NH446065	C1	3.27
NH4019	Lochan na Stairne	NH442057	C1	3.89
NH4023		NH404081	D	4.62
NH4029		NH421038	D	4.46
NH4030		NH421036	B	3.85
NH4032	Lochan Dearg Uillt	NH421031	C1	3.92
NH4037	Lochan nam Faoileag	NH426032	D	4.28
NH4038		NH430034	D	4.74
NH4039		NH434034	C1	3.96
NH4040		NH432030	B	2.95
NH4041	Lochan Carn a' Chuilinn	NH436037	D	4.62
NH4042	Loch Carn a' Chuilinn	NH440039	D	4.27
NH4104	Loch Kemp	NH468164	C2	3.85
NH4109	Loch nan Lann	NH441130	D	4.31
NH4110	Loch Knockie	NH455135	C2	4.37
NH4113	Loch Tarff	NH425100	C2	4.09
NH4114		NH413199	C2	3.31
NH4115	Loch a' Mheig	NH416199	B	3.41
NH4117		NH475163	B	4.15
NH4118		NH411199	C2	3.31
NH4246	Loch an t-Sionnaich	NH432216	C2	3.65
NH4249		NH452206	C1	3.65
NH4250	Loch a' Bhealaich	NH449207	D	5.21
NH4305	Lochan a' Bhathaich	NH408361	B	3.46
NH4308	Loch Garbh Iolachan	NH428358	C2	3.54

Site code	Site name	Grid Reference	Lake Group	PLEX Score
NH4309	Loch Neaty	NH432367	C2	3.85
NH4310	Loch Garbh Breac	NH434358	C2	3.57
NH4311	Loch Bruicheach	NH451365	C2	4.42
NH4313	Lochan an Tairt	NH445337	C2	3.94
NH4314		NH455343	B	4.26
NH4316	Loch Bad nan Earb	NH457340	B	3.57
NH4317	Loch nan Tunnag	NH456335	C2	4.47
NH4318	Loch Gorm	NH481334	C2	4.76
NH4320	Loch nam Bat	NH492332	C2	4.89
NH4321	Lochan Dubh	NH497336	B	3.40
NH4322		NH498331	C2	4.42
NH4323		NH484327	B	3.46
NH4324	Loch nam Faoileag	NH493325	C2	4.86
NH4325	Lochan an Torra Bhuidhe	NH485322	C2	5.03
NH4326		NH488316	C2	5.14
NH4327	Loch Meiklie	NH435301	D	4.87
NH4330		NH446318	B	4.38
NH4401	Loch nam Bonnach	NH480480	D	5.29
NH4402		NH435493	C2	3.67
NH4403	Loch Luaisgeach	NH438491	B	4.01
NH4405	Lochan Fada	NH427435	C2	4.23
NH4406	Loch na Cuile	NH431439	B	4.37
NH4410	Loch nan Gobhar	NH441440	B	3.94
NH4411		NH441436	C2	4.04
NH4412		NH438437	B	4.23
NH4416		NH437439	B	3.00
NH4501	Loch Garve	NH410596	D	4.94
NH4515	Loch Achilty	NH432566	C2	4.46
NH4522	Loch Kinellan	NH470575	D	5.75
NH4713	Loch Bealach Culaidh	NH448719	C2	4.23
NH4714	Loch nan Druidean	NH461713	C2	4.21
NH4802	Loch Chuinneag	NH494847	C2	4.26
NH5101	Loch Bran	NH509192	C2	3.94
NH5201	Loch Ceo Glais	NH590288	C2	4.73
NH5202	Loch a' Bhodaich	NH552245	B	4.23
NH5204	Loch Ness	NH515245	D	5.00
NH5205	Loch an Ordain	NH555240	C2	4.97
NH5206	Loch na Craoibhe-Beithe	NH544235	B	3.08
NH5207	Lochan Torr an Tuill	NH521228	B	5.04
NH5211	Loch Ruairidh	NH530213	B	4.34
NH5213	Loch Conagleann	NH585212	D	4.57
NH5214		NH591215	C2	4.02
NH5219		NH529213	B	4.34
NH5302	Loch Laide	NH546353	E	5.07
NH5401		NH555495	C2	4.96
NH5402		NH586416	C2	4.95
NH5404		NH580414	B	5.00
NH5501	Loch Ussie	NH505570	E	5.75

Site code	Site name	Grid Reference	Lake Group	PLEX Score
NH5511		NH580573	G	6.92
NH5513	Ord Loch	NH525505	D	6.06
NH5514		NH525503	G	6.32
NH5515		NH526506	H	7.12
NH5516		NH532502	A	4.42
NH5531		NH576535	A	1.54
NH5532		NH577535	A	2.31
NH5539		NH582534	B	4.15
NH5540		NH582533	B	4.15
NH5541		NH581532	A	2.31
NH5544		NH592540	A	2.56
NH5545		NH594542	A	2.98
NH5546		NH596543	A	1.54
NH5547		NH592539	B	3.75
NH5602		NH556687	B	4.76
NH5603	Breun Loch	NH556686	B	4.42
NH5701	Loch Morie	NH532759	C2	4.26
NH5801	Lochan a' Chairn	NH516843	C2	4.65
NH5901		NH561951	C2	4.65
NH5908	Loch a' Bhaid-bheithe	NH501902	C2	4.42
NH6201	Loch Ruthven	NH620276	E	5.32
NH6202		NH656283	B	3.65
NH6301		NH609395	D	6.23
NH6302	Loch Dochfour	NH607387	D	4.67
NH6303	Abban Water	NH600381	D	5.37
NH6306	Loch Ashie	NH630349	E	5.69
NH6307	Loch Bunachton	NH665350	D	5.69
NH6308	Loch na Curra	NH605324	D	4.42
NH6309	Lochan an Eoin Ruadha	NH611321	C2	4.58
NH6310	Loch Duntelchaig	NH624315	C2	4.42
NH6310.1	Loch nan Geadas	NH600306	B	4.56
NH6311	Loch a' Chlachain	NH655322	C2	3.57
NH6313	Loch Farr	NH685305	B	5.00
NH6403		NH636486	G	6.82
NH6501	Loch Culbokie	NH610590	B	4.23
NH6509	Loch Lundie	NH670504	E	5.58
NH6609		NH621614	G	6.54
NH6709	Loch Achnacloich	NH665736	I	6.85
NH6807	Loch Saine	NH696899	C2	4.91
NH6901	Loch Migdale	NH640908	E	5.16
NH6902		NH618923	C2	4.04
NH6903		NH617932	B	2.77
NH6904		NH617931	B	3.52
NH6905		NH612932	A	4.62
NH6906		NH609935	A	2.69
NH6907	Loch an Lagain	NH659956	C2	4.73
NH6908	Loch Buidhe	NH663983	C2	4.69
NH6909	Loch Laro	NH608994	C2	4.49

Site code	Site name	Grid Reference	Lake Group	PLEX Score
NH7005	Loch Gynack	NH744022	D	4.90
NH7007		NH798022	B	4.23
NH7009		NH798013	D	5.06
NH7010		NH798015	B	4.23
NH7308	Loch a' Chaorainn	NH755375	C2	4.15
NH7309	Loch Moy	NH775344	D	5.08
NH7317		NH782334	B	3.08
NH7501		NH732572	H	7.12
NH7602		NH719644	G	6.32
NH7711	Lochanan nan Tunnag	NH766761	D	5.85
NH7712	Black Loch	NH776757	G	6.54
NH7803	Black Pond	NH753836	B	5.81
NH7815	Loch Ospisdale	NH731889	G	7.45
NH7816		NH737881	I	8.85
NH7817	Lake Louise	NH735878	B	3.08
NH7818	Loch Evelix	NH740876	D	6.19
NH7823	Loch-an-treel	NH778899	D	5.93
NH7902	Loch as Airde	NH700902	C2	4.13
NH7903	Loch Lannsaidh	NH737945	C2	5.16
NH7904	Loch Laiogh	NH731960	C2	5.10
NH7905	Loch Ruagaidh	NH733994	B	3.72
NH8001	Loch Alvie	NH866096	D	4.88
NH8002	Loch Beag	NH862092	D	4.97
NH8009		NH802030	B	5.19
NH8010		NH803029	D	5.41
NH8012		NH809033	D	5.58
NH8013		NH812032	D	5.74
NH8014		NH813036	B	4.83
NH8015		NH815034	B	4.46
NH8017		NH817034	B	4.94
NH8018	Loch Insh	NH830043	D	5.27
NH8019		NH802017	D	5.21
NH8020		NH802016	G	6.12
NH8021		NH803016	D	5.21
NH8022		NH803013	B	4.84
NH8023	Lochan Geal	NH852058	C2	4.56
NH8026		NH874070	H	6.54
NH8030		NH881096	B	4.85
NH8031		NH897098	B	3.94
NH8032		NH899094	D	5.12
NH8034	Uath Lochan	NH836022	B	3.65
NH8035	Uath Lochan	NH837020	C2	3.77
NH8036	Uath Lochan	NH836018	C2	4.65
NH8037	Loch an Eilein	NH896076	C2	4.69
NH8038	Loch Gamhna	NH891068	C2	4.61
NH8041		NH836048	C2	4.23
NH8042		NH800023	B	5.72
NH8043		NH801015	D	5.47

Site code	Site name	Grid Reference	Lake Group	PLEX Score
NH8044		NH803028	D	5.77
NH8105		NH888124	C2	4.23
NH8107		NH890121	C2	4.49
NH8210		NH882261	A	1.54
NH8211		NH881263	A	1.54
NH8213		NH883261	B	4.17
NH8404	Loch of Boath	NH886451	D	5.26
NH8501	Loch Flemington	NH810520	G	6.98
NH8601		NH812691	I	7.69
NH8701	Loch Eye	NH832793	E	6.40
NH8705	Bayfield Loch	NH821718	E	6.06
NH8802	Loch nan Tunnag	NH833841	E	5.78
NH8803	Loch nan Tunnag	NH837837	E	5.78
NH8804	Loch na Muic	NH836841	E	5.45
NH8805	Red Stripe Loch	NH842830	C2	5.00
NH8806	Loch Preas an Uisge	NH847832	B	5.38
NH9006		NH952085	C2	3.85
NH9007		NH951086	C2	4.40
NH9008	Lochan nan Geadas	NH955088	D	5.00
NH9009	Loch Morlich	NH965093	D	4.89
NH9101		NH907165	D	5.70
NH9102		NH910177	C2	5.19
NH9103	Loch Vaa	NH914175	D	4.64
NH9105		NH943193	E	5.74
NH9108		NH905156	C2	5.32
NH9109	Loch nan Carraigean	NH903155	C2	4.88
NH9114	Loch Dallas	NH932159	D	5.24
NH9115		NH920138	B	3.08
NH9116	Loch Pityoulish	NH920136	C2	5.16
NH9120	Loch Mallachie	NH964173	C2	4.15
NH9121	Loch Garten	NH973180	C2	4.55
NH9216	Loch Mor	NH962255	G	6.24
NH9304	Lochindorb	NH971361	E	5.44
NH9427	Lochan Tutach	NH985404	D	4.27
NH9503		NH924583	C2	4.90
NH9504	Loch Loy	NH933587	E	5.85
NH9505	Cran Loch	NH946590	G	6.33
NJ0011	Loch Avon	NJ015025	C1	3.65
NJ0012		NJ018026	C1	3.65
NJ0016	Loch Etchachan	NJ006003	A	1.54
NJ0017		NJ012003	A	1.54
NJ0106	An Lochan Uaine	NJ001106	C2	4.23
NJ0203		NJ050281	D	6.79
NJ0304		NJ002395	A	1.54
NJ0305		NJ001393	A	1.54
NJ0410		NJ022416	A	1.54
NJ0411		NJ024417	A	1.54
NJ0504		NJ013551	G	8.85

Site code	Site name	Grid Reference	Lake Group	PLEX Score
NJ0602		NJ017607	D	5.98
NJ0603		NJ017606	D	5.85
NJ0609		NJ017608	H	6.67
NJ1202		NJ173290	D	5.16
NJ1507		NJ199590	H	7.69
NJ2301		NJ257396	D	5.38
NJ2502		NJ202591	I	7.43
NJ2611	Loch Spynie	NJ235665	I	7.55
NJ3405	Loch Park	NJ356430	G	6.76
NJ3504		NJ346591	F	5.51
NJ3606		NJ346637	G	6.81
NJ4008		NJ428076	D	5.88
NJ4016		NJ434011	B	2.56
NJ4018	Loch Davan	NJ441007	E	6.31
NJ4021	Braeroddach Loch	NJ481003	D	5.56
NJ4401		NJ417413	D	6.41
NJ5606		NJ588658	G	7.95
NJ6204		NJ692203	D	4.31
NJ6402	Haremoss Loch	NJ688445	H	7.69
NJ6503		NJ688516	G	7.56
NJ7009	Loch of Skene	NJ784075	I	7.60
NJ7107		NJ798160	G	6.36
NJ7303	Loch of Fyvie	NJ764387	G	7.19
NJ7502		NJ744548	G	6.79
NJ8002		NJ821041	G	7.38
NJ8305	Upper Lake	NJ878346	G	7.53
NJ8502		NJ845599	A	5.26
NJ8503		NJ849595	D	5.97
NJ9011	Loirston Loch	NJ938011	G	7.29
NJ9102	Bishops' Loch	NJ911142	D	6.29
NJ9104	Corby Loch	NJ924143	G	6.49
NJ9501		NJ944548	H	7.77
NJ9603		NJ980633	G	7.54
NK0202	Cotehill Loch	NK026292	G	7.24
NK0204	Sand Loch	NK033284	G	5.75
NK0303	Meikle Loch	NK029308	I	8.13
NK0503		NK039591	G	6.86
NK0601		NK000643	G	6.89
NL6901	Loch Tangusdale (Loch St. Clair)	NL647997	E	6.73
NL6902		NL631973	H	6.54
NL6903		NL634974	H	7.42
NL6905		NL631943	H	7.42
NL9301	Linne Thorramhuill	NL973391	B	4.19
NL9402	Loch Earblaig	NL945462	I	8.22
NL9403	Loch Bhasapoll	NL970470	I	7.15
NL9418		NL962431	C2	4.62
NL9419	Loch a' Phuill	NL960420	I	6.82

Site code	Site name	Grid Reference	Lake Group	PLEX Score
NL9422	Loch an Eilein	NL985436	E	6.32
NM0402	Loch Dubh a' Gharraidh Fail	NM026486	C2	3.93
NM0403		NM028485	C2	4.20
NM0406	Loch na Gile	NM027482	C2	4.27
NM0417	Loch Riaghain	NM033470	E	5.98
NM0419	Loch nam Braoileagan	NM011463	I	7.69
NM0426		NM072491	D	5.22
NM0427	Loch Monteich Mhoir	NM073492	C2	4.77
NM0428		NM076492	B	4.30
NM0433	Loch Dearg	NM075485	C2	5.06
NM1502		NM176580	E	5.64
NM1505		NM118524	E	7.10
NM1506	Lochan a' Chuirn	NM119523	E	5.82
NM1509	Loch nan Cinneachan	NM188565	C2	4.19
NM2201	Loch Staoiheig	NM265226	C2	4.63
NM2501	Loch Cliad	NM207586	C2	4.90
NM2505		NM231597	A	2.98
NM2506		NM232594	B	3.00
NM2507		NM230593	C2	3.38
NM2523	Loch nan Geadh	NM239582	C2	3.32
NM2524		NM238582	C2	3.29
NM2525		NM240598	C2	3.29
NM2526	Loch Urbhaig	NM230579	C2	3.93
NM2603	Loch a' Mhill Aird	NM231610	B	3.98
NM2605		NM234602	C2	3.15
NM2611	Loch Ronard	NM237608	C2	3.63
NM2615	Lochan a' Bhaigh	NM254631	B	3.73
NM2618		NM254616	C2	3.67
NM2619		NM254614	A	2.98
NM2622	Lochan Sagairt	NM250610	C2	4.40
NM3106	Loch Mor Ardalanish	NM363191	C2	3.73
NM3202	Loch Point na h-I	NM314227	E	5.66
NM3204	Loch an t-Suidhe	NM370214	B	4.19
NM3913	Long Loch	NM363985	C2	4.46
NM3916		NM376993	C2	3.68
NM3918	Priomh-lochs	NM369985	C2	4.20
NM3919	Loch Bealach Mhic Neill	NM376990	C2	3.94
NM3920		NM382991	C2	3.65
NM3922		NM379985	C2	3.80
NM3937	Loch Papadil	NM364920	D	4.89
NM4205	Loch Assapol	NM404205	E	4.83
NM4308		NM461338	C1	3.23
NM4309		NM460337	C1	3.00
NM4310		NM464337	C2	3.32
NM4311		NM462336	B	2.69
NM4312		NM463335	C2	3.74
NM4313		NM463334	C2	3.46
NM4314		NM457333	B	3.00
NM4404	Loch a' Gheal	NM415477	C2	4.57
NM4506	Lochan's Airde Beinn	NM473537	C2	3.65
NM4508	Loch Peallach/Loch Meadhoin/Loch Carnain an Amais	NM485530	C2	5.15
NM4601	Loch Grigadale	NM432668	B	5.10
NM4602		NM451683	C2	3.46
NM4603	Loch Caorach	NM433656	C2	3.31
NM4604	Lochan Druim na Claise	NM425644	C2	3.94
NM4605		NM437648	B	4.04
NM4606		NM438648	B	2.82
NM4608	Lochan an Aodainn	NM457660	C2	3.46
NM4609	Lochan na Crannaig	NM464657	C2	3.65
NM4610		NM446641	B	3.90
NM4616	Lochan Sron nan Sionnach	NM484656	C2	3.37
NM4617	Lochain Ghleann Locha	NM463639	C2	3.54
NM4618	Lochain Ghleann Locha	NM464637	C2	3.57
NM4702	Lochan an Dobhrain	NM478701	C2	4.12
NM4703	Lochan Dubh	NM490706	C2	3.88
NM4801		NM444868	C1	3.65
NM4804	Loch Beinn Tighe	NM450865	C2	4.04
NM4805		NM470875	A	2.88
NM4806		NM497886	D	5.67
NM4808		NM450860	C2	3.46
NM4813	Loch nam Ban Mora	NM455852	C2	4.23
NM4814		NM455849	C2	3.65
NM5233		NM590264	A	2.95
NM5234	Loch Fuaran	NM584268	C2	3.81
NM5235		NM584273	C2	3.21
NM5302	Loch Ba	NM570376	C2	4.59
NM5403		NM539417	C2	4.36
NM5404		NM552420	C2	3.99
NM5501	Lochan a' Ghurrabain	NM520538	G	6.15
NM5502		NM520530	A	3.27
NM5504	Lochan na Guailne Duibhe	NM528522	C2	4.86
NM5505		NM524522	B	3.78
NM5606	Lochan an Ime	NM556685	D	4.85
NM5609	Lochan nan Dearcag	NM559679	C2	4.33
NM5611		NM585696	C2	4.38
NM5616		NM587693	C2	3.74
NM5618	Lochain Beinne Brice	NM586688	C2	3.21
NM5627	Lochan a' Mhaidaidh Riabhaich	NM557656	D	4.29
NM5629	Lochan Poll an Dubhaidh	NM536645	D	4.51
NM5651	Lochan na Tuaidh	NM559677	C2	4.47
NM5704		NM582700	B	3.27
NM5705		NM585701	C2	4.07

97

Site code	Site name	Grid Reference	Lake Group	PLEX Score
NM6213	Loch Uisg	NM639251	D	4.92
NM6319	Loch Bearnach	NM689320	C2	3.87
NM6320	Crun Lochan	NM688301	B	4.39
NM6404		NM676405	B	4.33
NM6508	Loch Doire nam Mart	NM661525	C2	4.42
NM6511	Lochan Beinn Iadain	NM694535	B	4.31
NM6512		NM698532	C2	3.90
NM6513	Loch Arienas	NM683512	C2	4.68
NM6602	Lochan na Creige Duibhe	NM641667	C2	4.23
NM6616	Loch Loisgte	NM629649	C2	4.10
NM6617		NM633657	C2	3.85
NM6619	Lochan Sligneach	NM641653	C2	3.08
NM6623		NM643652	C2	3.27
NM6628		NM636625	A	2.31
NM6629	Loch Laga	NM640634	C2	4.23
NM6649	Lochain Ruighe a' Bhainne	NM691664	C2	3.73
NM6650	Lochain Ruighe a' Bhainne	NM691662	B	3.35
NM6656		NM692652	B	3.75
NM6657	Lochan na Bracha	NM699652	C2	3.42
NM6701	Loch Dubh	NM661774	B	3.00
NM6702		NM651739	C2	4.12
NM6703	Loch na Bairness	NM656757	C2	3.74
NM6708	Loch na Draipe	NM675749	C2	3.57
NM6720		NM685744	C2	3.85
NM6723		NM687745	B	2.69
NM6724		NM685742	B	2.69
NM6725	Loch Ard a' Phuill	NM690745	C2	3.59
NM6728		NM691740	C2	3.65
NM6729		NM672713	C2	3.88
NM6730		NM671711	C2	3.65
NM6732		NM673708	C2	4.00
NM6734	Loch Blain	NM673706	C2	4.78
NM6809	Loch Torr a' Bheithe	NM648842	C2	3.73
NM6811	Loch nan Eala	NM668859	B	4.81
NM6819	Loch Dubh	NM672848	B	4.23
NM6822		NM631850	B	2.98
NM6825		NM678855	B	3.08
NM6826		NM678854	B	3.08
NM6901	Lochan Doilead	NM676946	B	3.01
NM6902	Loch an Nostarie	NM690955	C2	4.23
NM6904	Loch a' Ghille Ghobaich	NM586941	C2	3.88
NM6905		NM581941	B	3.59
NM6906	Lochan a Mheadhoin	NM693948	C2	3.97
NM6907	Loch a Bhada Dharaich	NM695946	C2	4.09
NM7005	Loch Fada	NM787043	C2	4.78
NM7006	Loch Mhic Mhairtein	NM783032	D	5.03
NM7114		NM747144	C2	4.23
NM7115	Ballachuan Loch	NM761153	J	8.27

Site code	Site name	Grid Reference	Lake Group	PLEX Score
NM7301	Lochan an Doire Dharaich	NM717337	B	3.81
NM7302	Loch a' Ghleannain	NM727313	C2	4.12
NM7404	Loch Tearnait	NM748469	C2	4.42
NM7407	Loch na Sula Bige	NM770467	C2	3.51
NM7408		NM767466	B	2.95
NM7517	Lochanan Dubha	NM704541	C2	4.23
NM7518	Lochanan Dubha	NM703528	C2	3.67
NM7601		NM714686	A	3.85
NM7602		NM713678	A	1.54
NM7603		NM741677	A	1.54
NM7609		NM751658	C1	3.08
NM7610		NM753658	C2	3.35
NM7611		NM753659	A	2.31
NM7612		NM751657	C1	3.08
NM7613		NM766658	C1	3.85
NM7614	Dubh Lochain	NM714616	B	3.57
NM7615		NM718617	B	3.01
NM7619	Lochan Dhonnachaidh	NM705605	C2	3.85
NM7620		NM703604	C2	3.19
NM7622	Loch Doilet	NM798679	C2	3.91
NM7631		NM787688	B	3.21
NM7632		NM783688	C2	3.52
NM7737		NM718737	C2	3.08
NM7738		NM722737	C2	3.54
NM7743	Loch nam Paitean	NM725740	C2	3.85
NM7744		NM730736	C2	3.46
NM7745	Loch Dearg	NM735742	C2	3.85
NM7752		NM763736	C2	4.40
NM7757	Lochan na Creige	NM758722	C2	3.55
NM7758	Lochan na Creige	NM753717	C2	3.88
NM7759		NM754716	C2	3.77
NM7761		NM761720	C1	3.35
NM7762		NM762724	B	3.00
NM7771	Loch nan Lochan	NM745726	C2	3.85
NM7772		NM716739	C2	3.65
NM7773	Upper Loch an Sligeanach	NM715742	C2	4.10
NM7802	Lochan na Ba Glaise	NM713888	C2	3.85
NM7803	Lochan Donn	NM710886	C2	4.20
NM7804	Lochan Fada	NM708883	C2	3.85
NM7805		NM708881	C2	4.65
NM7811	Lochan a' Bhealaich	NM711876	C2	4.23
NM7838	Loch an Fhearainn Duibh	NM712821	C2	3.74
NM7839	Loch Doir a' Ghearrain	NM725817	C2	4.23
NM7840	Loch Mama	NM754852	C2	4.23
NM7845		NM756845	C2	3.21
NM7846	Lochan na Creige Duibhe*	NM755844	C2	3.65
NM7847		NM756844	C2	3.00
NM7848		NM758844	C1	3.65

Site code	Site name	Grid Reference	Lake Group	PLEX Score
NM7852		NM736821	C2	3.88
NM7854	Lochan a' Bhealaich	NM741822	C2	4.06
NM7866	Loch Eilt	NM795823	C2	4.13
NM7869	Loch na Creige Duibhe	NM765850	C2	4.35
NM7870	Loch a' Choire Riabhaich	NM719877	C2	4.23
NM7875		NM709880	C2	3.94
NM7920	Lochan Stole	NM745935	C2	3.71
NM7921	Lochan Innis Eanruig	NM736930	C2	3.85
NM7922		NM750935	C2	3.37
NM7923		NM751935	C2	3.19
NM7925	Lochan a' Chuirn Duibh	NM744927	C2	3.65
NM7943	Lochan Ropach	NM746930	C2	3.54
NM8003	Loch na Beiste	NM813058	D	5.45
NM8009	Lochan Fearphorm	NM835036	C2	5.25
NM8012	Loch Ederline	NM867025	D	5.73
NM8014		NM863018	D	5.66
NM8113	Loch a' Phearsain	NM854137	D	5.28
NM8121	Loch na Curraigh	NM864130	C2	4.35
NM8122		NM860126	E	5.29
NM8124	Loch a' Mhinn	NM864127	C2	4.33
NM8201	Lochan na Circe	NM809285	C2	4.27
NM8212	Dubh Loch	NM800207	D	5.54
NM8213	Loch Seil	NM805204	D	6.24
NM8301	Loch Fiart	NM809376	I	6.52
NM8304	Lochan Dubh	NM866320	C2	5.24
NM8404	Loch Baile a' Ghobhainn	NM859425	I	7.00
NM8509	Loch Uisge	NM806549	C2	4.42
NM8605	Lochan Feith nan Loagh	NM839670	C2	3.35
NM8619		NM848666	C1	2.95
NM8706		NM868797	C2	3.21
NM8707		NM869797	C2	3.21
NM8715	Lochan nan Sleubhaich	NM874798	C2	3.61
NM8823		NM832839	C2	3.46
NM8824	Lochan Stob a Glas-chairn	NM830836	C2	4.10
NM8825		NM832837	C2	3.19
NM8837	Loch Shiel	NM899800	C2	4.10
NM8841		NM831837	A	2.31
NM9006	Dubh Loch	NM911017	C2	4.62
NM9008		NM913017	C2	4.07
NM9010	Fincharn Loch	NM930035	C2	3.55
NM9011	Dubh Loch	NM937036	C2	3.85
NM9018		NM944039	C2	4.23
NM9019	Loch Gaineanhach	NM912007	C2	4.99
NM9020	Loch nan Eilean	NM945036	C2	4.35
NM9024	Loch Geoidh	NM950035	C2	3.90
NM9030	Loch nan Losgann	NM954031	C2	3.90
NM9031		NM951027	C2	3.85
NM9032	Dubh Loch	NM942025	C2	3.94

Site code	Site name	Grid Reference	Lake Group	PLEX Score
NM9037		NM926012	D	4.62
NM9067	Loch Leacann	NM997032	D	4.71
NM9070	Loch Tunnaig	NM915015	C2	4.06
NM9103	Loch na Sreinge	NM926170	D	5.06
NM9106	Loch Avich	NM934145	D	4.84
NM9213		NM937241	D	4.73
NM9214	Lochan Sonachan	NM939244	C2	4.78
NM9215		NM942241	B	3.65
NM9218		NM961236	B	3.08
NM9219	Loch a' Bharraom	NM965241	C2	4.15
NM9222		NM958229	C2	4.16
NM9223	Sior Loch	NM962299	C2	4.73
NM9224	Sior Loch	NM970232	C2	4.50
NM9309	Lochan na Beithe	NM916352	C2	4.23
NM9310	Lochan nan Rath	NM920353	C1	4.46
NM9406		NM918443	B	3.92
NM9501	Lochan Doire a' Bhraghaid	NM925585	C2	3.57
NM9503	Lochan Torr an Fhamhair	NM931582	C2	4.12
NM9504	Lochan na Criche	NM922573	C2	3.65
NM9614	Loch nan Gabhar	NM969632	C2	4.62
NM9806	Lochan Port na Creige	NM908802	C2	3.90
NN0053	Steallaire ban Loch	NN070092	C2	4.26
NN0057		NN076086	D	5.43
NN0101		NN017183	C2	4.29
NN0142	Loch Awe	NN004164	D	5.46
NN0209	Loch Tromlee	NN043251	E	5.79
NN0310		NN023314	D	4.72
NN0402	Loch Baile Mhic Chailein	NN022474	C2	4.75
NN0504		NN032578	A	2.95
NN0505		NN031577	B	3.08
NN0506		NN028574	B	2.98
NN0507		NN031572	B	2.95
NN0508		NN030571	C1	3.27
NN0606		NN005640	A	1.54
NN0608		NN006637	D	4.55
NN0911	Lochan Dubh	NN060955	C2	3.72
NN0912		NN057951	C2	3.19
NN1113	Dubh Loch	NN116111	J	7.31
NN1116		NN175105	D	5.48
NN1201		NN137277	B	4.73
NN1202		NN138277	B	4.73
NN1320		NN154307	C1	3.97
NN1502	Loch Achtriochtan	NN142567	C2	4.51
NN1503		NN152552	A	1.54
NN1504		NN153552	A	1.54
NN1505		NN153551	A	2.31
NN1710	Lochan Meall an t-Suidhe	NN143727	C2	3.90
NN1711		NN162718	A	1.54

Site code	Site name	Grid Reference	Lake Group	PLEX Score
NN1712		NN163718	A	1.54
NN1802	Loch Lochy	NN190870	D	5.00
NN1908	Lochan Ceann Caol Glas Bheinn	NN113965	C1	3.08
NN1909		NN113964	A	2.31
NN1916		NN105950	A	2.31
NN1930		NN105950	A	1.54
NN2001		NN230086	C2	4.46
NN2002	Loch Restil	NN229080	C2	4.54
NN2153		NN276158	A	2.31
NN2156	Lochan Srath Dubh-Uisge	NN281157	C2	3.85
NN2164		NN277156	B	2.95
NN2206	Lochan a' Mhadaidh	NN267217	C2	3.85
NN2322	Lochan Coire Thoraidh	NN218314	C1	3.92
NN2404		NN268481	A	2.69
NN2410		NN285439	A	3.33
NN2414		NN289436	B	3.00
NN2415		NN287482	B	2.69
NN2419	Loch Dochard	NN211418	C2	4.13
NN2421	Loch Buidhe	NN296482	C2	4.00
NN2422		NN285471	A	2.56
NN2511	Lochan na Fola	NN209560	B	2.98
NN2514		NN295548	A	1.54
NN2516	Lochan Mathair Eite	NN288543	C2	3.42
NN2518		NN293543	A	1.54
NN2534		NN285508	A	2.95
NN2615		NN223656	C1	3.65
NN2616		NN225654	C2	4.51
NN2617	Loch Eilde Beag	NN255654	C2	4.12
NN2619	Loch Eilde Mor	NN253639	C2	4.52
NN2623	Loch Chiarain	NN291636	C2	3.94
NN2624		NN247628	A	2.31
NN3003		NN391098	C2	4.23
NN3101	Geal Loch	NN318164	C2	4.70
NN3102	Dubh Lochan	NN324166	C2	4.30
NN3204	Loch Oss	NN300252	C2	3.89
NN3303	Lochan na Bi	NN308312	C2	4.52
NN3401	Lochan na Stainge	NN301492	C2	4.04
NN3407		NN306485	B	2.69
NN3408	Loch Ba	NN318497	C2	3.92
NN3413	Lochan na h-Achlaise	NN310479	C2	3.50
NN3524		NN312537	A	2.31
NN3525		NN315537	B	3.54
NN3531	Lochan Gaineamhach	NN303535	C2	3.85
NN3538		NN327524	A	2.56
NN3556		NN318511	B	3.46
NN3602	Loch Ossian	NN390680	C2	4.02
NN3607		NN367664	A	1.54

Site code	Site name	Grid Reference	Lake Group	PLEX Score
NN3608		NN365662	A	2.31
NN3610		NN367662	A	2.31
NN3611	Feur Lochan	NN360660	C2	3.40
NN3615		NN363658	C2	3.90
NN3616	Loch na Sgeallaig	NN368658	C2	4.07
NN4001	Loch Katrine	NN435093	D	4.74
NN4002	Loch Chon	NN4205	C2	4.02
NN4007	Loch Ard	NN4702	C2	4.83
NN4010	Loch Dhu	NN4303	C2	4.13
NN4103	Loch Doine	NN4619	C2	5.19
NN4104	Loch Voil	NN5020	C2	4.55
NN4207	Loch Dochart	NN4025	D	5.51
NN4214	Loch Iubhair	NN4227	C2	4.62
NN4313	Lochan Learg nan Lunn	NN448385	C2	3.80
NN4716	Loch Coire Cheap	NN480755	D	4.85
NN4722	Loch an Sgoir	NN489750	D	4.51
NN4805		NN418881	A	3.85
NN4806		NN420883	A	1.54
NN4807	Lochan Uaine	NN420881	A	1.54
NN4809	Lochan a Choire	NN437882	C2	4.40
NN5002	Loch Achray	NN514063	D	4.98
NN5016	Lake of Menteith	NN576004	D	6.02
NN5103	Loch Lubnaig	NN581132	D	5.07
NN5201	Lochan an Eireannaich	NN514242	E	5.55
NN5202	Lochan Lairig Cheile	NN5527	C2	3.90
NN5408	Lochan na Lairige	NN597401	B	3.08
NN5505	Loch Rannoch	NN584574	C2	4.18
NN5702	Loch Pattack	NN539790	C2	4.18
NN6008	Muir Dam	NN660023	H	6.00
NN6301	Loch Tay	NN674374	D	5.68
NN6402	Lochan nan Cat	NN645426	E	5.73
NN6404	Lochan nan Uan	NN648425	E	5.56
NN6608	Loch Con	NN687679	C2	3.99
NN6903	Loch Glas-Choire	NN630908	C2	4.58
NN6905	Lochain Uvie	NN675956	B	4.38
NN6906		NN678958	C2	4.71
NN6908		NN674948	B	4.81
NN6910	Loch Etteridge	NN691931	C2	4.10
NN6911		NN672946	B	4.42
NN7001	Loch Mahaick	NN7006	D	6.18
NN7006	Loch Watston	NN711002	G	7.61
NN7403		NN734465	G	6.54
NN7505	Loch Kinardochy	NN775552	C2	4.94
NN7603	Maud Lochan	NN725659	C2	4.70
NN7701	Loch an Duin	NN722799	E	5.04
NN7809	Loch Bhrodainn	NN747831	D	4.87
NN7903		NN708976	B	4.81
NN7905		NN749990	C2	4.50

Site code	Site name	Grid Reference	Lake Group	PLEX Score
NR8803	Daill Loch	NR813898	D	4.62
NR8904	Lochan Add	NR862976	C2	5.07
NR8905	Loch Leathan	NR873983	C2	4.48
NR8907	Lochan an Torrnalaigh	NR855954	C2	4.66
NR9201	Loch Cnoc an Loch	NR934286	D	4.67
NR9202		NR946228	D	5.63
NR9203		NR949222	D	6.37
NR9204		NR947218	H	7.31
NR9318		NR952314	A	1.54
NR9321		NR967395	A	1.54
NR9401		NR909491	J	4.23
NR9802		NR934821	D	5.14
NR9901	Lochan Anama	NR904988	C2	3.73
NR9902	Lochan Breac-Liath	NR913989	C2	3.80
NR9907	Loch Glashan	NR917933	H	6.15
NS0201	Urie Loch	NS002280	D	5.10
NS0202	Loch na Leirg	NS020271	B	2.98
NS0203	Loch Garbad	NS018238	C2	4.81
NS0204		NS012297	H	7.12
NS0503	Loch Quien	NS064593	E	6.46
NS0603	Greenan Loch	NS067640	G	7.07
NS0611	Loch Fad	NS077616	I	6.69
NS0613	Kirk Dam	NS082630	G	6.73
NS0702	Bull Loch	NS008731	C2	3.80
NS0801	Loch Tarsan	NS077841	C1	4.81
NS0901	Garbhallt Lochain	NS029945	C2	5.14
NS0902		NS051940	B	2.95
NS1001		NS191004	G	8.08
NS1003		NS190004	G	6.62
NS1511		NS172572	D	5.81
NS1512		NS168568	D	5.54
NS1515		NS167566	B	2.56
NS1516		NS167567	B	2.31
NS1517		NS171567	A	2.31
NS1526		NS173573	H	6.54
NS1601		NS118691	D	4.62
NS1602		NS118690	D	5.93
NS1603		NS117689	G	6.79
NS1604		NS104635	G	6.23
NS1704	Loch Loskin	NS168786	D	5.15
NS1904	Loch Eck	NS137915	D	4.91
NS2001	Swan Lake	NS224096	G	6.92
NS2002		NS228093	G	8.46
NS2021		NS236099	G	6.73
NS2029		NS229093	G	8.46
NS2102		NS277150	D	4.90
NS2409	Stevenston or Ashgrove Loch	NS274443	G	6.71

Site code	Site name	Grid Reference	Lake Group	PLEX Score
NS2501	Haylie Reservoir	NS216580	D	6.71
NS2602	Kelly Reservoir	NS223684	D	4.69
NS2603	Blackfield Loch	NS212677	A	1.54
NS2703	Coves Reservoir	NS248764	D	6.52
NS2707		NS239734	D	5.98
NS2711		NS240729	D	5.77
NS2807	Lindowan Reservoir	NS241815	D	4.57
NS2901	Corran Lochan	NS216952	C2	4.27
NS3001		NS310094	G	7.02
NS3002		NS313096	G	6.79
NS3112	Martnaham Loch	NS394174	G	7.18
NS3114	Snipe Loch	NS385173	G	7.46
NS3201		NS341246	G	7.07
NS3301		NS307377	I	7.79
NS3309		NS358336	B	5.43
NS3311		NS395327	G	8.18
NS3416		NS326422	I	7.29
NS3504	Castle Semple Loch	NS363590	G	6.89
NS3505	Barr Loch	NS352576	G	7.22
NS3607		NS367698	D	5.66
NS3608	Knapps Loch	NS365684	E	6.88
NS3722		NS394712	B	6.13
NS3723		NS395715	D	5.86
NS3804		NS305838	D	5.28
NS3805		NS307838	B	4.21
NS3902	Dubh Lochan	NS375962	C2	4.42
NS4004	Bogton Loch	NS468055	D	6.01
NS4102	Belston Loch	NS475168	D	6.11
NS4104		NS400170	H	7.12
NS4204		NS436218	I	8.40
NS4303		NS479369	H	7.69
NS4402	Burnfoot Reservoir	NS451448	G	6.92
NS4503	Caplaw Dam	NS434587	D	6.37
NS4504	Loch Libo	NS433556	G	6.64
NS4602		NS414668	G	6.65
NS4708	Lily Loch	NS472778	D	4.72
NS4709	Burncrooks Reservoir	NS484791	H	6.69
NS4803		NS444878	G	7.08
NS4904		NS469960	B	5.14
NS5002	Loch Muck	NS512007	D	4.85
NS5105	Black Loch	NS590162	G	6.74
NS5210		NS530206	G	7.19
NS5301		NS532365	G	6.82
NS5405		NS563417	A	1.54
NS5507	Brother Loch	NS506527	D	6.12
NS5508	Little Loch	NS503521	B	5.50
NS5704	Carbeth Loch	NS533792	D	6.12
NS5711	Drumbrock Loch	NS549783	C2	3.99

Site code	Site name	Grid Reference	Lake Group	PLEX Score	Site code	Site name	Grid Reference	Lake Group	PLEX Score
NS5728		NS548728	F	8.05	NS8617	Riven Loch	NS830621	A	1.54
NS5730	Dougalston Loch	NS561737	G	7.20	NS8619		NS847623	A	5.34
NS5733		NS543714	G	7.13	NS8701		NS832795	E	6.35
NS5735	Bardowie Loch	NS578735	G	7.55	NS8809		NS875825	G	7.95
NS5804		NS543805	I	8.12	NS9001		NS917082	D	6.47
NS5905	Loch Macanrie	NS561993	D	5.59	NS9101		NS938123	D	5.37
NS6001		NS614062	A	1.54	NS9201		NS946271	G	7.60
NS6105		NS605134	G	7.95	NS9302		NS942372	H	6.54
NS6113		NS605135	G	8.08	NS9303	Lochlyoch Reservoir	NS932356	D	5.88
NS6202		NS670258	D	6.41	NS9307		NS978358	H	6.79
NS6402	Parkfield Loch	NS644420	A	1.54	NS9319		NS937387	H	7.12
NS6504		NS678516	G	6.83	NS9401		NS928459	D	6.46
NS6605	Hogganfield Loch	NS642672	I	7.86	NS9405	Red Loch	NS955473	G	7.31
NS6724	Gadloch	NS648710	I	7.66	NS9408	White Loch	NS961471	H	7.69
NS6803	Loch Walton	NS665866	D	6.18	NS9413		NS972437	G	7.09
NS7001		NS778097	H	6.92	NS9414		NS920438	G	8.25
NS7101		NS789120	H	6.54	NS9418		NS933462	H	7.12
NS7202		NS710263	G	6.63	NS9419		NS933463	G	6.86
NS7203		NS706261	B	2.95	NS9508		NS965534	H	7.12
NS7205	Glenbuck Loch	NS757287	I	7.35	NS9616		NS921617	D	5.27
NS7304		NS798362	D	5.30	NS9711	Lochcote Reservoir	NS976736	I	6.81
NS7402		NS768435	D	5.96	NS9905	Gartmorn Dam	NS919941	I	7.24
NS7504		NS713577	H	7.12	NT0008		NT078015	F	6.77
NS7525		NS762522	G	7.31	NT0101		NT032155	D	6.64
NS7605	Woodend Loch	NS704667	I	6.97	NT0201	Cowgill Lower Reservoir	NT007291	D	6.11
NS7607	Lochend Loch	NS705662	I	7.68	NT0301		NT001341	H	7.12
NS7703		NS707772	G	7.71	NT0304		NT076395	G	6.54
NS7808		NS741829	D	6.06	NT0412		NT071421	D	6.47
NS7902		NS735973	A	1.54	NT0603		NT003625	G	6.20
NS8004	Waddels Loch	NS866026	G	6.92	NT0702	Linlithgow Loch	NT002776	I	8.18
NS8101		NS889139	A	1.54	NT0705		NT010742	G	6.82
NS8102		NS891141	A	1.54	NT0804	Town Loch	NT098892	I	8.12
NS8106		NS803120	D	5.91	NT0905	Black Loch	NT075961	E	5.73
NS8201		NS841206	D	5.58	NT0908	Loch Glow	NT085956	D	5.53
NS8307		NS834315	D	6.23	NT1003		NT110017	G	6.81
NS8309		NS844320	G	7.17	NT1102	Loch Skeen	NT171165	D	4.44
NS8310		NS839314	I	7.34	NT1103		NT175160	A	1.54
NS8315		NS870362	G	7.07	NT1304		NT122346	G	6.85
NS8401		NS826497	H	5.96	NT1502	North Esk Reservoir	NT154581	D	6.35
NS8414		NS871489	G	6.22	NT1506	Slipperfield Loch	NT135505	B	4.90
NS8417		NS873488	G	7.44	NT1511		NT136507	A	1.54
NS8427		NS899430	G	7.39	NT1512		NT136506	A	1.54
NS8429		NS845482	H	7.69	NT1608	Threipmuir Reservoir	NT163635	D	6.43
NS8518		NS866512	D	6.50	NT1705		NT161783	G	8.46
NS8519		NS865511	D	5.38	NT1814		NT187877	I	8.30
NS8520		NS865510	H	7.12	NT1912	Loch Fitty	NT121914	I	7.25
NS8604		NS801653	I	7.32	NT2001	Moodlaw Loch	NT294073	A	1.54
NS8606	Black Loch	NS861697	D	5.32	NT2101	Loch of the Lowes	NT237198	E	6.30

Site code	Site name	Grid Reference	Lake Group	PLEX Score
NT2201	St Mary's Loch	NT247227	D	5.97
NT2304	Loch Eddy	NT281308	D	6.24
NT2405		NT210438	G	6.86
NT2514	Gladhouse Reservoir	NT298534	D	6.31
NT2602	Clubbiedean Reservoir	NT200668	D	6.12
NT2604		NT251693	D	5.96
NT2707		NT276748	G	7.95
NT2710		NT253708	I	8.40
NT2711	Dunsapie Loch	NT280731	I	8.33
NT2712	Duddingston Loch	NT281724	I	7.83
NT2807	Kinghorn Loch	NT258873	I	7.91
NT2907	Camilla Loch	NT220915	I	7.74
NT3101	Clearburn Loch	NT341155	D	6.09
NT3106	Kingside Loch	NT339133	G	7.27
NT3110	Alemoor Reservoir	NT398151	D	6.55
NT3201		NT394299	H	6.15
NT3303		NT302333	G	7.50
NT3511		NT309542	G	6.75
NT3601		NT337686	H	7.69
NT4009		NT493088	H	7.12
NT4110	Branxholme Wester Loch	NT422110	I	6.13
NT4111	Branxholme Easter Loch	NT433117	I	7.55
NT4113	Williestruther Loch	NT491114	I	7.26
NT4210		NT498290	G	6.24
NT4213		NT497277	D	6.08
NT4214	Akermoor Loch	NT406209	I	6.98
NT4217		NT498272	G	7.44
NT4220	Essenside Loch	NT461207	I	7.29
NT4306		NT485333	G	7.24
NT4402		NT470447	G	7.44
NT4501	Fala Flow	NT427585	H	7.69
NT4601		NT455690	G	6.00
NT4703		NT457787	I	8.42
NT4801		NT467810	G	6.89
NT5001		NT559079	G	7.10
NT5116		NT540121	G	7.12
NT5117		NT539119	G	6.67
NT5201		NT502291	G	6.58
NT5212	Newhall Loch	NT560270	G	7.46
NT5306	Faldonside Loch	NT504328	G	7.76
NT5312		NT523307	G	7.65
NT5313		NT540317	G	6.26
NT5405		NT584436	I	7.12
NT5611	Danskine Loch	NT565676	G	7.01
NT5801		NT503843	G	8.16
NT6001		NT668036	A	1.54
NT6202		NT615261	G	8.85
NT6205		NT641259	B	5.38
NT6312		NT640346	G	7.26
NT6407		NT689460	H	7.38
NT6502		NT654503	I	8.04
NT6603		NT681619	G	7.69
NT6704		NT627732	G	7.62
NT6801		NT607834	I	8.37
NT6802		NT608834	I	8.55
NT7102	Whitton Loch	NT744197	G	7.79
NT7203	Wooden Loch	NT708253	G	7.97
NT7401	Hule Moss	NT713490	H	7.69
NT7402		NT718492	H	7.69
NT7505	Hen Poo	NT778546	G	7.90
NT7603		NT782673	G	6.73
NT7703		NT764718	B	5.96
NT8101		NT810166	I	7.24
NT8202	Yetholm Loch	NT803279	I	8.33
NT8303	Hoselaw Loch	NT808317	F	7.25
NT8404	Hirsel Lake	NT824402	G	8.55
NT8504		NT808550	H	7.69
NT8607	Coldingham Loch	NT894685	I	7.62
NT8701		NT806706	H	7.44
NT9602	Mire Loch	NT911686	G	8.26
NW9603		NW992631	D	5.37
NW9703		NW988717	H	8.04
NX0404		NX087425	I	7.81
NX0510		NX091559	H	6.54
NX0601	Loch Connell	NX017681	G	7.77
NX0701	Kilantringan Loch	NX090790	D	6.12
NX0801		NX098800	G	6.81
NX0802		NX081825	J	7.31
NX1301		NX113393	G	7.29
NX1403		NX115453	G	7.22
NX1515		NX189590	G	7.69
NX1606	Black Loch	NX111616	D	6.60
NX1701		NX141761	A	1.54
NX1801	Loch Melemon	NX118858	G	7.21
NX1802		NX124862	G	6.83
NX1805		NX165877	G	8.85
NX1902	Loch of Lochton	NX174924	B	5.06
NX1905		NX198904	B	5.88
NX2504	Dernaglar Loch	NX263581	D	5.06
NX2601	Eldrig Loch	NX252693	D	5.26
NX2704	Long Loch	NX238763	D	4.65
NX2706	Black Loch	NX240762	C2	3.92
NX2806	Near Eyes Stanks	NX250845	B	4.90
NX2902	Penwhapple Reservoir	NX261974	D	5.99
NX2907	Dinmurchie Loch	NX289925	A	1.54
NX3301		NX396371	D	6.53

Site code	Site name	Grid Reference	Lake Group	PLEX Score
NX3408		NX363415	G	7.22
NX3502	Black Loch	NX301545	C2	4.54
NX3608		NX310617	D	5.68
NX3609		NX308611	F	6.20
NX3715	Dow Lochs	NX374719	D	4.87
NX3716	Dow Lochs	NX365717	B	4.62
NX3806	Kirriereoch Loch	NX363865	C1	4.58
NX3901	Linfern Loch	NX367980	D	4.96
NX4311		NX466359	G	6.80
NX4414		NX478423	G	7.85
NX4503		NX402541	G	6.88
NX4607	Bruntis Loch	NX446653	C2	4.19
NX4709	Black Loch	NX496728	C2	4.01
NX4823	Loch Valley	NX444817	C2	3.35
NX4826	Loch Narroch	NX452815	C1	3.23
NX4913	Loch Doon	NX496970	D	5.12
NX5401		NX583496	I	7.84
NX5516		NX581505	G	6.24
NX5517		NX581503	G	6.31
NX5601	Coo Lochans	NX507693	A	2.31
NX5602	Coo Lochans	NX508692	A	2.31
NX5701	Clatteringshaws Loch	NX545769	D	5.23
NX5803	Loch Dungeon	NX526844	B	3.40
NX5815		NX528848	B	3.59
NX5905		NX509989	D	4.57
NX5906		NX510991	D	4.57
NX6402		NX616491	H	6.78
NX6403		NX618491	H	7.44
NX6418		NX678454	G	7.98
NX6532		NX663509	H	7.42
NX6533		NX660507	H	6.74
NX6614	Bargatton Loch	NX691617	D	5.91
NX6616	Woodhall Loch	NX671674	D	5.56
NX6715	Loch Ken	NX657729	D	5.67
NX6718	Stroan Loch	NX644703	C2	4.77
NX6802	McKay's Loch	NX691896	D	5.46
NX6902	Kendoon Loch	NX611904	D	6.00
NX7407		NX766486	D	6.70
NX7505	Jordieland Loch	NX713537	D	5.54
NX7616	Carlingwark Loch	NX763613	G	7.62
NX7702	Lowes Lochs	NX707786	B	3.56
NX7803	Loch Urr	NX760845	D	5.45
NX7901	Stroanshalloch Loch	NX700906	G	5.62
NX7902	Troston Loch	NX700903	C1	3.72
NX8401	Loch Mackie	NX808489	D	5.54
NX8511	White Loch	NX864548	G	5.93
NX8604	Edingham Loch	NX837633	F	6.73
NX8608		NX841601	B	4.52
NX8708	Milton Loch	NX839715	G	7.25
NX8812		NX882815	D	6.19
NX8911	Morton Loch	NX891992	D	6.53
NX9501		NX901583	G	7.44
NX9601	Loch Arthur	NX904688	D	5.67
NX9604	Loch Kindar	NX967642	D	5.71
NX9703	Lochaber Loch	NX921703	D	6.06
NX9821		NX995805	G	6.51
NX9907	Loch Ettrick	NX944937	D	5.63
NY0602		NY045665	H	6.54
NY0707		NY050738	B	3.08
NY0821	Castle Loch	NY086814	G	7.76
NY0903		NY044912	G	7.06
NY1601		NY139689	G	7.33
NY1604		NY141661	G	8.85
NY1801		NY101872	F	5.96
NY2602		NY293691	G	7.22
NY2701	Purdomestone Reservoir	NY214775	G	6.93
NY2802	Winterhope Reservoir	NY275824	D	6.22
NY2901	Black Esk Reservoir	NY204967	D	5.75
NY3701		NY323702	H	6.54

Site code	Site name	Grid Reference	Lake Group	PLEX Score
NN8303	Loch Freuchie	NN864375	D	5.43
NN8408	Loch na Craige	NN883456	C2	4.83
NN8611		NN863633	B	4.55
NN9204		NN935285	G	7.76
NN9403	Loch Kennard	NN907460	C2	4.52
NN9509		NN966533	B	5.41
NN9901	Loch Einich	NN914989	C2	3.93
NO0303		NO054384	F	5.00
NO0413	Loch of Craiglush	NO042442	D	5.14
NO0509	Loch Ordie	NO033500	D	5.47
NO0603		NO039639	D	6.67
NO0902	Lochan Uaine	NO026985	C1	3.08
NO0905		NO037955	B	3.01
NO1005	Loch Leven	NO144015	I	7.46
NO1308	Kings Myre	NO112362	I	7.32
NO1402	Loch of Clunie	NO113441	D	5.82
NO1403	Loch of Drumellie or Marlee Loch	NO141442	D	6.19
NO1415	Black Loch	NO174427	C2	5.38
NO1609	Drumore Loch	NO165607	I	7.11
NO1701	Loch Vrotachan	NO122785	E	5.43
NO1805	Loch Phadruig	NO176861	D	4.74
NO1806	Loch Callater	NO184839	D	4.74
NO1809	Loch Kander	NO190808	D	4.62
NO1902		NO139907	C2	5.38
NO2006	Ballo Reservoir	NO224049	E	6.07
NO2102	Lochmill Loch	NO222162	I	7.93
NO2206		NO279295	G	8.16
NO2312	Laird's Loch	NO258356	E	5.99
NO2319	Lochindores	NO270356	G	6.54
NO2405	Monk Myre	NO208427	G	8.05
NO2505	Loch of Lintrathen	NO275546	G	6.94
NO2802	Sandy Loch	NO228864	C1	4.13
NO2803		NO230862	D	5.77
NO2805		NO225854	C1	2.88
NO2806		NO226856	D	3.85
NO2808		NO254864	A	2.95
NO2809		NO253864	A	2.95
NO2810		NO253863	A	2.95
NO2811	Lochnagar	NO252860	C1	3.65
NO2812	Dubh Loch	NO238827	C1	3.85
NO2814	Loch Muick	NO288828	C1	4.00
NO2901		NO267937	G	7.24
NO3003	Carriston Reservoir	NO326037	I	7.83
NO3111		NO322112	G	6.77
NO3210		NO344218	G	7.22
NO3601		NO376606	B	4.49
NO3903		NO393984	C2	3.35

Site code	Site name	Grid Reference	Lake Group	PLEX Score
NO4006	Kilconquhar Loch	NO487017	I	7.51
NO4114	Cameron Reservoir	NO471112	I	7.34
NO4205	Morton Lochs	NO462266	G	7.32
NO4206	Morton Lochs	NO462261	E	6.43
NO4407		NO403415	G	7.29
NO4607		NO456633	D	5.77
NO4902	Loch Kinord	NO441993	D	5.86
NO4903		NO454988	B	5.43
NO4906		NO484961	B	4.87
NO4907		NO476955	C1	4.23
NO4909		NO457988	B	4.90
NO5001		NO509064	I	8.46
NO5104		NO577147	I	8.85
NO5105		NO547133	B	5.08
NO5301	Monikie Reservoirs	NO507383	I	7.27
NO5409	Crombie Reservoir	NO522406	G	6.91
NO5605		NO573667	G	7.82
NO5901	Loch of Aboyne	NO536998	D	6.22
NO5902		NO540995	G	6.31
NO6301		NO616395	G	7.33
NO6510	Nicholl's Loch	NO648536	A	2.56
NO6607	Dun's Dish	NO647609	I	7.63
NO6707	Loch Saugh	NO676788	G	6.42
NO6904		NO606910	A	5.19
NO6905		NO608911	C1	4.71
NO7603		NO709604	G	7.97
NO8903		NO857938	G	8.85
NR1501		NR166540	B	3.00
NR1502		NR165538	B	3.59
NR2401	Lower Glenastle Loch	NR294450	E	4.91
NR2503		NR246597	A	4.10
NR2504		NR249593	B	4.81
NR2515	Lochan na Nigheadaireachd	NR281558	D	5.85
NR2521		NR289563	A	2.56
NR2601	Loch Corr	NR225695	C2	4.75
NR2607		NR248692	B	4.08
NR2608		NR247691	A	2.98
NR2610		NR247690	A	2.31
NR2620	Loch Gorm	NR230657	C2	4.62
NR2706	Ardnave Loch	NR284727	E	6.38
NR3404	Loch nan Gabhar	NR337483	C2	5.77
NR3409	Glenastle Loch	NR300447	C2	4.67
NR3417	Loch nan Gillean	NR308431	C2	3.97
NR3418	Loch Ard Achadh	NR312427	E	5.66
NR3419	Loch Kinnabus	NR301422	E	6.35
NR3506	Loch Tallant	NR336579	B	4.44
NR3507	Loch Tallant	NR335578	B	4.62
NR3508	Loch Airigh Dhaibhaidh	NR315555	C2	3.81

101

Site code	Site name	Grid Reference	Lake Group	PLEX Score
NR3514		NR318549	A	1.54
NR3544	Loch Eighinn	NR330504	C2	4.34
NR3631	Loch Skerrols	NR342638	E	6.24
NR3633	Loch Finlaggan	NR386676	D	5.37
NR3637	Loch Bharradail	NR393635	C2	4.76
NR3802	Lochan Chille-moire	NR360888	B	4.33
NR3803		NR354877	B	5.34
NR3902		NR370958	A	2.69
NR3904		NR388969	B	5.15
NR3905		NR374952	C2	5.26
NR3907		NR357934	E	6.55
NR3908		NR394965	G	6.89
NR3909		NR394959	C2	5.30
NR3910	Loch Fada	NR383955	C2	5.00
NR3912	Loch an Squid	NR382947	B	4.04
NR3913		NR382946	B	3.59
NR3914		NR388945	B	4.06
NR4406	Loch nan Digl	NR431481	B	4.58
NR4535	Loch Tallant	NR448505	B	3.38
NR4602	Loch nan Cadhan	NR404668	E	5.47
NR4605	Loch Allan	NR424678	D	5.63
NR4606	Loch Ballygrant	NR405662	E	6.40
NR4607	Loch Lossit	NR408652	E	5.73
NR4608	Loch Fada	NR408636	C2	5.21
NR4616	Loch a' Bhaile Mhargaidh	NR495669	D	4.58
NR4702	Loch Staoisha	NR406712	C2	4.42
NR4703	Ardnahoe Loch	NR421715	C2	3.97
NR4708	Loch nam Ban	NR419705	C2	4.42
NR4901	Lochan a' Bhraghad	NR415983	B	3.78
NR5603	Loch a' Mhuilinn	NR518669	C2	4.67
NR5721		NR524768	C1	3.46
NR5722		NR525768	C1	3.46
NR5731	Loch na Fudarlaich Beag	NR528765	C1	3.46
NR5732	Loch na Fudarlaich	NR532763	C1	3.69
NR5748		NR537759	C1	3.40
NR5750		NR543757	A	1.54
NR5751	Loch na Cloiche	NR539753	C2	3.50
NR58122		NR558843	A	1.54
NR58123		NR558842	C1	2.95
NR58124		NR560841	C1	3.27
NR58125		NR561840	A	2.88
NR58126		NR562840	A	2.31
NR58127		NR563841	A	2.31
NR58129		NR569841	A	2.31
NR6101	Loch an t-Olais	NR635192	D	5.93
NR6102	Killypole Loch	NR648188	D	6.14
NR6201	Tangy Loch	NR695280	E	6.37
NR6203		NR653220	H	7.69
NR6501	Upper Loch	NR638502	C2	5.42
NR6701		NR692788	H	6.54
NR6702		NR696779	D	5.70
NR6818	Loch Cathar nan Eun	NR631869	B	3.16
NR6821		NR604838	B	3.01
NR6822	Loch a' Mhuilinn	NR606837	B	3.52
NR6936	Loch Doire na h-Achlaise	NR655957	C2	3.55
NR6937	Lochan Carn Thearlaich	NR661962	B	3.19
NR6938	Loch nan Eilean	NR661965	C2	3.73
NR7201	Skeroblin Loch	NR703262	C2	5.34
NR7204	Aucha Lochy	NR726226	D	6.21
NR7205	Knockruan Loch	NR733227	C2	5.05
NR7301	Dubh Loch	NR712398	A	1.54
NR7420	Loch Mor	NR730404	C2	3.57
NR7421	Loch Beag	NR734401	A	2.56
NR7501	Loch nan Gad	NR782571	E	5.00
NR7502	Loch na Beiste	NR765546	C2	4.51
NR7503	Loch Ciaran	NR774541	C2	4.62
NR7601	Lochan Chaorann	NR737685	B	3.37
NR7602	Lochan Eun	NR745688	B	4.36
NR7603	Loch nan Torran	NR756688	C2	4.51
NR7614	Loch a' Bharra Leathain	NR771622	B	3.37
NR7615	Loch Cill' an Aonghais	NR776617	E	5.38
NR7801		NR727863	C2	5.28
NR7803	Loch Coille-Bharr	NR780899	D	5.49
NR7805	Lochan Taynish	NR740857	C2	4.88
NR7806		NR703828	E	5.53
NR7809	Loch McKay	NR798885	C2	4.49
NR7811	Loch nam Ban	NR757860	C2	3.46
NR7812	Loch a' Mhuilinn	NR762834	C2	3.54
NR7813	Loch nam Breac Buidhe	NR761829	C2	3.77
NR7814		NR765833	C2	3.85
NR7815	Loch na h-Uamhaidh	NR767836	C2	3.65
NR7816	Loch na Creige Crainde	NR764827	C2	4.33
NR7818	Loch na Sgratha	NR769821	C2	3.54
NR7834	Loch an Fhir-Mhaoil	NR744892	B	3.80
NR7835		NR712820	J	5.77
NR7836		NR708820	H	7.00
NR7837		NR750896	B	2.31
NR7902		NR750902	B	2.98
NR7903	Loch Barnluasgan	NR791911	D	5.52
NR7905	Loch Linne	NR797910	C2	4.23
NR8301		NR892337	D	5.49
NR8702	Lochan Amna	NR816791	C2	4.21
NR8703	Loch Fuar-Bheinne	NR811784	C2	4.23
NR8708	Meall Mhor Loch	NR851737	C1	3.69
NR8801	Lochan Duin	NR804897	C2	4.46
NR8802	Loch na Bric	NR802891	F	5.00